George Riley

The beauties of the creation

A new moral system of natural history. Second American Edition

George Riley

The beauties of the creation
A new moral system of natural history. Second American Edition

ISBN/EAN: 9783337202118

Printed in Europe, USA, Canada, Australia, Japan

Cover: Foto ©berggeist007 / pixelio.de

More available books at **www.hansebooks.com**

THE

BEAUTIES of the CREATION;

OR, A NEW

MORAL SYSTEM

OF

NATURAL HISTORY;

DISPLAYED IN THE

MOST SINGULAR, CURIOUS, AND BEAUTIFUL

QUADRUPEDS, BIRDS, INSECTS, TREES, AND FLOWERS:

Defigned to infpire Youth with Humanity towards the Brute
Creation, and bring them early acquainted with the
wonderful Works of the Divine Creator.

SECOND AMERICAN EDITION.

" *Who can this field of Miracles furvey,*
" *And not with* GALEN *all in rapture fay,*
" *Behold a* GOD! *adore him, and obey.*"
BLACKMORE on the Creation.

PHILADELPHIA:

Printed by WILLIAM YOUNG, Bookfeller, N°. 52,
Second, corner of Chefnut-ftreet.

1796.

PREFACE.

*NATURAL HISTORY, in its general
sense, comprehending the whole produce of the
creation, as consisting of beasts, birds, fishes,
insects, reptiles, flowers, plants, stones, fossils,
and minerals, it was impossible to include, in a
single volume like the following, even the names
of the different articles: we were therefore o-
bliged to make a selection of a part, which we
considered the most curious, interesting, and
worthy the attention of the pupil studying that
science of nature.*

*In the progress of the work, those beasts,
birds, insects, and flowers, are particularly
described, that are distinguished by any peculiar
characteristics of beauty, utility, curiosity, or
medicinal virtue.*

*It has been our endeavour to trace more those
grand outlines of sublime wonders that elevate
the heart to the Creator, than to descend to*

the

*the minute inveſtigation of a mere ſpeculatiſt,
for, in the work of the ingenious Blackmore,*

" Who can this field of miracles ſurvey,
" And not with GALEN* all in rapture ſay,
" Behold a GOD! adore him, and obey!"

*This engaging ſubject, much as it is neg-
lected, is, of all others, the moſt neceſſary to
finiſh a polite education. It ſoftens and humanizes the mind imperceptibly; for it leads
us to this ſublime truth—That nothing is created in vain; and teaches us, that a knowledge
of God is the moſt noble, and ſhould therefore
be the ultimate object of all our purſuits. This
divine ſcience is therefore the only means by
which we can know ourſelves, and be grateful
for thoſe beings that are created for our uſe,
ſupport and protection.*

*We have been more anxious to vindicate
the dignity of nature, than to debaſe it with
puerile reſearches. Whenever any grand deviation was obſervable in one beaſt or bird
from another, we made free to ſearch for the
final cauſe, independent of former opinions,*

---

* GALEN was profeſſedly an atheiſt, until he providentially
ſaw an human ſkeleton, which, conſidering attentively, with
regard to the wiſdom diſplayed in its ſtructure, was the immediate cauſe of his not only believing in a God, but becoming
a moſt zealous profeſſor of religion.

however

however fanctioned by authority, when they happened not to be congenial with our own sentiments. To trace the final caufes, or the reafons of the difference in the various claffes of birds and beafts, is the firft and moft effential object to purfue in the ftudy of nature. To look for differences, as fome have done, only to gratify a prepoffeffion for novelty, without improving the mind or amending the heart, is to turn natural hiftory into a raree-fhow, inftead of adopting it as a fcience.

To avoid that tedious detail of defcription which tires by its famenefs, and confufes by its intricacy, we have fpecified only thofe characteriftics that were effential to notice, in order to be able to diftinguifh one animal from another: but, in this, the peculiar beauties are more particularly noticed than any deviation of colour or form, that had no quality to recommend it to our attention.

With refpect to the arrangement, we have endeavoured to prefent it as fyftematically to our readers, as an abridgement could poffibly admit. That the ftudent might know of what fpecies every bird and beaft was, which this volume contains, they are defcribed in the order of their refpective claffes. Whenever there were more of a fpecies than the limits of the work would admit of being defcribed, they are fpecified by name, according to the moft accurate naturalifts.

A. 2.

It being the desire of the proprietor of this work to render it as complete as in his power, he has taken all possible care to give the most correct descriptions of the different beasts, birds, insects, and flowers. He hopes therefore his endeavours to render it instructive and interesting, will be received as a small token of that respect he has for the public patronage he now ventures to solicit.

INDEX.

INDEX.

I N S E C T S.

T R E E S.

F L O W E R S.

EXPLA-

EXPLANATION of SCIENTIFIC TERMS,

OCCURRING

IN THE FOLLOWING SUBJECTS OF

NATURAL HISTORY.

Abdomen, the belly.

Aurelia, the state of the insect, while changing from the worm to the moth, fly, or butterfly.

Apterous, without wings.

Antennæ, horns or feelers.

Chrysalis, the same as aurelia.

Crustaceous, covered with a shell, or a substance similar to a shell.

Capitulum, a little head.

Elytra, the cases of the wings.

Forceps, the forked tail of an insect.

Genus, several beings agreeing in one common character.

Hymenoptera insecta, insects having four membraneous wings.

Individual, a being considered separately from others of the same species or kind.

Larva, the worm or caterpillar.

Lobe, a division or distinct part.

Lapidoptera insecta, insects having four wings.

Membraneous, consisting of a fibrous web.

Maxilla, the jaws.

Nympha, see Aurelia.

Neuroptera insecta, insects with four transparent wings.

Palpi, spiral tongues.

Porrected, stretched out.

Reticulated, formed like net-work.

Scarabæus, the beetle.

Setaceous, covered with bristles.

Species, a common nature, by which several individuals are distinguished.

Spine, a thorn.

Thorax, the breast.

Vermicula, the nature of the insect before it begins its transformation.

ADDRESS.

A D D R E S S.

A NEW MORAL SYSTEM OF

NATURAL HISTORY,

COMPOSES the following volume, treating of quadrupeds, birds, infects, and flowers. This engaging fubject, much as it is neglected, is of all others, the moft neceffary to finifh a polite education, imperceptibly, as it foftens and humanizes the mind, while, by leading us to this fublime truth, that nothing is created in vain, we obtain, what ought to be the ultimate object of all our purfuits, a knowledge of GOD, of ourfelves, and of the beings he has formed for our ufe, fupport, and protection.

Such are the general outlines of the prefent work, now fubmitted to the judgment of parents and the guardians of youth; a work whofe fole object is to form an happy union of inftruction and amufement. In a word, to render what has been perverted into an irkfome burthen to the youthful mind, an agreeable and a rational paftime.

The proprietor has only to obferve that the fuccefs, which has already attended his endeavours to promote the love of virtue and knowledge in the above work, has far exceeded his moft fanguine expectations. He begs leave, in gratitude, to declare, that no trouble, no expence, have been fpared to render the prefent enlarged edition, a work of lafting utility to the rifing generation, and an agreeable ufeful pocket companion, to all who wifh to acquire a knowledge of the above important and interefting fubjects, by

The public's obliged and devoted fervant,

GEORGE RILEY.

FROM

REVIEWS, &c.

WE acknowledge with real regret that the present system of female education is too imperfect and confined; and we earnestly look and long for that happy period, when the mind's of Nature's fairest works will not be shackled by an improper course of education, and when the gratification received by solid instruction will effectually induce them to bestow on their children the valuable acquisition of a well informed mind.

To remove, in a great measure, the force of the above argument, is, we think, one of the leading purposes of this valuable publication. Here useful knowledge is collected, classified, and expressed in the most concise, simple, and easy manner.

Its object is to serve as a key to open the mind to extensive views of the natural and moral worlds, and to lead youth to admire the wisdom, and adore the goodness of GOD; the first grand and valuable principle of every thing virtuous and praise worthy. We therefore heartily recommend it to the use, not only of children, but of parents themselves, as a publication which, in every respect, does the head and heart of the editor the highest honor.

The editor of the Biographical Magazine observes, that — The subjects of these useful, moral, and elegant (this work was published in 2 vols.) volumes, are treated with great taste, ingenuity of observation, and morality of principle.

NATURAL

NATURAL HISTORY.

QUADRUPEDS.

THEIR GENERAL NATURE.

QUADRUPEDS, after MAN, in natural hiſtory, require the next attention, and for the following reaſons. Being of ſimilar ſtructure with ourſelves, having inſtincts and properties ſuperior to all other parts of animated nature, affording great aſſiſtance to man, and ſometimes exerciſing the greateſt hoſtilities, muſt render them the moſt intereſting part of the creation, and claim the firſt attention of the naturaliſt.

Similitude to man.—Like us they are elevated above the birds, by their young being produced alive ; above the claſs of fiſhes, by breathing through the lungs ; above inſects, by blood circulating through their veins ; and moſtly above all parts of the creation, by being partly or entirely covered with hair. Since quadrupeds ſo nearly approach us in

B   animal

animal perfection, how little reafon have we
to be vain of our corporeal qualities !

Figure.—The heads of quadrupeds are generally adapted to their mode of living.   In
fome it is fharp, to enable them to turn up
the earth, where they find their food depofited ; in others, it is long, in order to afford
room for the olfactory nerves ; in many, it
is fhort and thick, to ftrengthen the jaw, and
qualify it for combat.   Their legs and feet
are entirely formed to the nature and exigencies of the animal.   When the body is heavy, the legs are thick and ftrong ; when it is
light, they are active and flender.   Thofe
that feed on fifh, are made for fwimming, by
having webbed feet : thofe that prey upon
animals, are provided with claws which they
can draw and fheath at pleafure ; but the
more peaceable and domeftic animals are generally furnifhed with hoofs, which, being
more neceffary for defence than attack, enable them to traverfe the immenfe tracts which
they are deftined to pafs over, either to ferve
man, fearch for food, or avoid hoftility.

Difpofition for prey.—Beafts of prey feldom devour each other.   Nothing but extreme hunger can induce them to commit
this outrage againft nature ; and, when they
are obliged to feck fuch fubfiftence, the weakeft affords to the ftrongeft but a difagreeable
repaft.   In fuch cafes, the deer or goat is
what they particularly feek after, which they
either take by purfuit or furprife.

Natural

Natural fagacity.—In countries uninhabited by man, fome animals have been found in a kind of civil fociety, where they feemed united in mutual friendfhip and benevolence : but no fooner does man intrude upon their haunts, than their bond of fociety is diffolved, and every animal feeks fafety in folitude.

Cloathing of animals.—In the colder climates they are covered with a fur, which preferves them from the inclemencies of the weather ; in the more temperate, they have fhort, and in the warmer climates, they have fcarcely any hair upon their bodies. Thus we perceive that they are provided with cloathing, according to the nature of their fituation.

Ferocity.—Where men are the moft barbarous, animals are the moft ferocious. Thofe produced in climates of extreme heat, poffefs a nature fo favage, that they are fcarcely ever tamed.

Food.—The place, as well as the nature of their food, is adapted to the fize and fpecies of the animal. Thofe feeding in vallies are generally larger than thofe that feek their food on mountains. In warm climates, their plenteous and nutritive food renders them remarkable for bulk. Milk is their firft aliment.

Produce.—Beafts that are large, ufelefs, and formidable, produce but few at a time, while thofe that are fmall, ferviceable, and inoffenfive, are more prolific. This feems

to

to be adapted with the moſt admirable proportion ; for, were the ſmaller and weaker to
have leſs offspring, their race might be deſtroyed, by being ſo frequently made the prey
of ſtronger animals.

Courage.—In defence of their young, no
danger or terror can drive animals from their
protection. Such as have force, and ſubſiſt by
rapine, are moſt formidable in their ferocious
courage.

Generation.—Each ſpecies of quadrupeds
bring forth their young at the time when nature moſt plenteouſly affords them their reſpective nutriment. Thoſe animals which
hoard up proviſions for the winter, produce
their young in January, by which time they
are enabled to collect ſufficient ſubſiſtence for
their offspring. Quadrupeds which are called
oviparous, from being hatched from eggs, ſuch
as the crocodile, turtle, &c. are the moſt prolific, being no ſooner freed from the ſhell than
they attain their utmoſt ſtate of animal perfection.

Every ſpecies of animal has its peculiar
cry, by which they diſtinguiſh each other,
and communicate the general expreſſions of
their paſſions, as fear, joy, deſire &c. Thus
has the all wiſe, bountiful, and divine creator, in his infinite wiſdom, formed a race
of animals for the uſe of mankind, and
granted us dominion over them, which ſhould
never be exerciſed but with the greateſt humanity.

The

The H O R S E.

OF all quadrupeds, the horfe is the moft generous, ferviceable, and beautiful. There is none to which man is more indebted. Wild horfes herd together, in affemblies of five or fix hundred, and depute one as a centinel to guard the reft while fleeping. Arabia is the moft famous for this animal in its wild ftate., But the Englifh horfe excels all in fize, utility, and fwiftnefs. It is longer lived than the Barb, and more hardy than the Perfian. The famous Childers was ,fo fleet as to run a mile in a ininute. The Englifh hunters are allowed to be the moft ufeful horfes in the world. To give a ' defcription of this well-known animal is unneceffary, as there is fcarcely a country in which he is not to be found. Spain, Italy, Denmark, Germany, Hungary, Holland, Flanders, France, Crete, Morocco, Turkey, Perfia, India, China, Tartary, and Arabia, abound with various fpecies of them, which differ according to the foil and climate of the country. But the general received opinion is, that the native clime of this noble animal is Arabia, to which all the countries above mentioned are indebted for the different breeds of horfes they poffefs.

Its difpofition to war caufed it to be confecrated to Mars, the god of battle.

Of

Of their hides are made collars, and all kinds of harnefs; their manes and tails are ufed in perukes, lines for angling, covering for chair-feats, cords, floor-cloths, and a variety of other articles.

Although they are endowed with vaft ftrength, and great powers, yet they feldom exert either to the prejudice of their mafters; on the contrary, they will cheerfully encounter the greateft fatigues for their benefit. They fear and love the human race, and are of a very benevolent difpofition. And yet, notwithftanding all the good qualities of this noble and generous animal, when he is fo enfeebled by age, and worn down by the fevere drudgery of his lordly mafter, as to be incapable of contributing any longer to his pleafure, his ambition, or his avarice, he is (as if ingratitude was peculiar to the human fpecies) fold for fcarcely the worth of his bridle. In this ftate of lamentable exiftence, he is configned to the cruel treatment of fome inhuman wretch, who chaftifes him for that weaknefs incident to his old age, or which he has acquired in the fervitude of his former mafter, and thus tortures the remnant of his life, which fhould, were it only for paft fervices, be cherifhed with the moft tender care and attention.

Such is the ftrength of the Englifh draught horfe, that in London they have been feen to draw three tons weight. In Yorkfhire, the pack-

pack-horfes ufually carry a burthen of 420lb. over the higheft hills.

The A S S.

THIS animal refembles the horfe very nearly in form, but, being of a dif-tinct fpecies, in a ftate of nature it is entirely different. It is found wild in the deferts of Lydia and Numidia, where it is. caught with traps. Of their fkins, fhagreen leather, and other valuable articles are manufactured. The plantain is their favourite vegetable. Their fcent is fo acute, that they are capable of fmelling their driver or owner at a great diftance, and will even diftinguifh him in a croud. In proportion to his fize, he is ftronger than the horfe, and fupported with much lefs care and fuftenance. In fome countries they are fo large, that in Spain a jack-afs is frequently feen fifteen hands high. Of all animals covered with hair, the afs is the leaft fubject to vermin. His period of exiftence is from twenty to twenty-five years ; and, although he can endure much more fatigue and hardfhip than a horfe, he has much lefs fleep. It is related of this animal, that he will never ftir if he be blinded.

The afs was originally imported into Amecica by the Spaniards, who now hunt them for their diverfion.

In

In his natural ftate, he is fleet, fierce, and formidable, but when domefticated, he is the moft gentle of all animals, and affumes a patience and fubmiffion even humbler than his fituation. He is very temperate in eating, and contents himfelf with the refufe of the vegetable creation. As to drink, he is extremely delicate, for he will flake his thirft at none but the cleareft brooks, and thofe to which he is moft accuftomed. When young, he is fprightly, and tolerably handfome ; but age deprives him, as well as all other parts of animated nature, of thofe qualities ; he then becomes flow, ftupid, and obftinate. The fheafs goes eleven months with young, and never produces more than one at a time.

The ingenious author of the *Spectacle de la Nature*; obferves in fubftance, that though he is not poffeffed of very fhining qualities, yet what he enjoys are very folid ; that the want of a noble air hath its compenfation in a mild and modeft countenance ; that his pace is uniform, and, although he is not extraordinary fwift, he purfues his journey a long while without intermiffion ; and that he is perfectly well contented with the firft thiftle that prefents itfelf in his way ; in fhort, that this indefatigable animal, without expence or pride, replenifhes our cities and villages with all forts of commodities.

With refpect to their general difpofition, the fame author informs us, " That the afs refembles

" refembles thofe people who are naturally
" heavy and pacific, whofe underftanding and
" capacity are limited to hufbandry or com-
" merce, who proceed in the fame track with-
" out difcompofure, and complete, with a fe-
" rious and pofitive air, whatever they have
" once undertaken."

The medicinal virtues of affes-milk, in re-
ftoring health and vigour to our debilitated
conftitutions, might alone entitle this harmlefs
and inoffenfive animal to a kinder return, than
it generally experiences from their inhuman
and ungrateful mafters.

The ZEBRA.

THIS animal is the moft wild and beautiful in nature, and is principally found in
the fouthern parts of Africa. It is faid to fur-
pafs. all others in fwiftnefs, and even ftands
better and firmer upon its legs than the horfe.
There was one in England that would eat
bread, meat, and tobacco. It differs from
the wild afs, with which it has been frequently confounded, in the defcription given
of it by fome naturalifts. In fhape, it more
refembles the mule, than the horfe or the
afs : it is lefs than the former, and longer
than the latter : its ears are longer than
thofe of the horfe, and fhorter than thofe
of the afs : it has a large head, a ftraight
back

back, well-placed legs, and tufted tail. The
fkin is clofe and fmooth, and the hind-quar-
ters are round and well formed. The male
is white and brown ; the female white and
black. The colours are fo regularly ftriped,
that they appear to be painted, and refemble
fo many ribbons laid over its body ; fo that,
at a fmall diftance, the Zebra appears to have
been dreffed by art, inftead of being fo admirably adorned by nature.

The M U L E.

THIS animal is bred between a horfe
and a fhe-afs, or a jack-afs and a mare.
In Spain, where they are ufed to draw people of the firft diftinction, they are frequently
fold at fifty or fixty guineas each. The common Mule is very healthy, and lives about
thirty years.

RUMINATING.

RUMINATING ANIMALS

ARE fuch as are diftinguifhed for chew-ing the cud, and being the moft mild and eafily tamed. The ferocious or the carnivo-rous kinds, feek their food in gloomy folitude; but thefe range together in herds, and the very meaneft of them unite in defence of each other. The food of ruminating animals being eafily procured, they feem more indolent, and lefs artful than the carnivorous kinds, or thofe which feed on flefh.

The BULL, OX, and COW.

OF all ruminating animals thefe are firft in rank, both with refpect to fize, beauty, and fervice. Many of our Englifh peafants have only a cow, from which they obtain a livelihood. Cows improve the pafture which affords them their nourifhment. Their age is calculated by their horns and teeth. Of all creatures, this animal is moft affected by difference of foil, which being luxuriant, increafes their growth to a confiderable fize, while in more fterile countries they are pro-portionally diminutive. In Great-Britain, the ox is the only horned animal that will apply

his

ſtrength to the ſervice of mankind. The ox, in particular, will grow to a prodigious ſize, an extraordinary inſtance of which is at this time to be ſeen in London , he was bred at Gedney, in the county of Lincoln, and is allowed by judges to be much the larg-eſt and fatteſt ox ever ſeen in England ; his beef and tallow alone being computed to weigh 350 ſtone, or 2800 pounds weight.

There is no part of this animal without utility ; the blood, fat, marrow, hide, horns, hoofs, milk, cream, whey, urine, liver, gall, ſpleen, and bones, have each their particular qualities. The hide, when tanned, is manu-factured into boots, ſhoes, and various other accommodations in life ; vellum and goldbeat-er's ſkin are alſo obtained from theſe animals ; the hair, mixed with lime, is uſed to cement our buildings : combs, knife-handles, boxes, buttons, drinking veſſels, &c. are made of their horns, which are alſo uſed as antidotes to poiſon, the plague, and ſmall-pox : glue is made from the chips of their hoofs, and the parings of the raw hides. Their bones are an excellent ſubſtitute for ivory ; and their feet afford an oil, ſo generally known under the name of *neats' foot oil*, that it needs no deſcription here. The blood is an excellent manure for fruit trees, and the chief ingredi-ent of Pruſſian-blue : the gall, liver, ſpleen, and urine, are uſed in Medicine. Milk, cheeſe, cream, and butter, are too common to require particular mention. The fleſh is

of

of two forts, namely, veal and beef, which, being dreffed various ways, is calculated to invigorate. the weak, fupport the laborious, and gratify the voluptuous.

The urus, or Wild Bull, is generally found in Lithuania, a province of Poland.

There are other fpecies of the cow-kind, fuch as the Bifon, Bonafus, Zebu, Beevehog, Buffalo, and Siberian cow.

The B U F F A L O.

T H E Buffalo, being more clumfy, is lefs beautiful than the cow. His fkin is alfo harder, thicker, blacker, and thinner of hair ; his flefh is hard, black, and difagreeable, both to the tafte and fmell : the milk, though a-bundant, is not fo good as that which the cow affords ; in the warm countries, however, it is ufed to make cheefe and butter. The hide, from its thicknefs and impenetrability, is dref-fed, and forms an article called *buff leather*, after his name.

Two of thefe animals, yoked together, will draw more than four ftrong horfes. When purfued, they will often fwim over the larg-eft rivers with great facility. They are found wild in many parts of Africa and Afia, and are likewife very common in Italy, from whence they were brought into Lombardy, A. D. 591. They grow to twice the fize of

C our

our largeſt oxen, and their horns are ſo large, that a pair is to be ſeen in the muſeum, which meaſure ſix feet, ſix inches, and a half in length, weigh forty two pounds, and hold ten quarts of water in their vacuities. Ariſtotle, very properly, calls theſe creatures wild oxen.

In the weſtern parts of Florida, on this ſide the Miſſiſippi, the buffalo is hunted after the following manner : the hunters range themſelves in four lines, forming a very large ſquare; they then ſet fire to the graſs, which is long and dry ; the animals draw cloſer together, as the fire runs along the lines, of which they are much afraid, and naturally fly from it, until they get quite cloſe together, they then attack them briſkly, ſeldom ſuffering any to eſcape. At theſe hunting-matches they generally kill from a thouſand to fifteen hundred of theſe animals.

The buffalo, like other animals that feed on graſs, is inoffenſive when undiſturbed ; but when wounded, or even fired at, their fury is ungovernable.

In India, there is a ſmaller kind of buffalo, which they make uſe of to draw their coaches.

In the northern parts of America there is another animal, larger than the ox, which has ſhort black hair, horns, a large beard, and a head ſo covered with hair that he makes a moſt formidable appearance.

ANIMALS

ANIMALS of the SHEEP and GOAT KIND.

ALTHOUGH this fpecies comprehends 'many animals of a fimilar nature, they differ with regard to their bodies, horns, food, and covering.

The utility, and inoffenfive nature of thefe animals, is a proof that they have been long reclaimed from their wild ftate, and adapted to domeftic purpofes. They both appear to require protection from man, whom they reward with the greateft favours ; they feem indeed, to court his fociety. Though the fheep is moft ferviceable, the goat has more attachment and fenfibility. In the earlieft ages, the goat appears to have been the greater favourite, and ftill continues fo amongft the poor. But the fheep has Jong been the principal object of human care and attention; we fhall therefore begin with

---

The SHEEP.

THIS animal, in its tame ftate, is the moft harmlefs and defencelefs. When wild, it is faid to be of vaft fwiftnefs and only found in great flocks. As foon as they are attacked, they form a ring, into the centre of which the

the ewes fall, where they are defended by
the rams in the moft vigorous manner. The
woolly fheep is only found in Europe, and
fome of the temperate provinces of Afia.
When fat, it is aukward in its motions, eafily fatigued, and frequently finks under the
weight of its own corpulence, and rich
fleeces. There is no part of this admirable
animal, but what has its particular ufe.

When two rams meet, they engage very
fiercely. Every ewe knows its lamb, and e-
very lamb the bleating of its ewe, even a-
midft thoufands. In England, they chiefly
feed on downs, in paftures, young fpring-
ing corn lands, or turnip fields ; but the
downs have, by long experience, been found
to prove by far the moft beneficial, on account of the air and drynefs of foil, no animal being fo fubject to the rot, as fheep, if
fed on marfhy land. The whole flock of
ewes, wethers, and lambs, are fheared once
in a year. Wethers have generally more
and better wool than the ewes. Such is their
utility in agriculture, that an hundred fheep
will manure eight acres of ground.

In Iceland they have a fpecies of this animal, called Many-horned fheep; they are of
a dark brown colour, and under the outward
coat of hair, have a fine, fhort, foft fur, refembling wool.

In Spain, the fheep produce a wool, fuperior to that of any other country. It is of
fo excellent a quality, that our hatters and
clothiers

clothiers are obliged to purchase it at a very great price, in order to enable them to manufacture some of their estimable articles.

The great utility of sheep to Great-Britain may be seen by the following moderate calculation of fleece-wool annually produced by their growth.

According to the calculation of Young, in his *Six months tour*, there are 466,532 packs of wool manufactured in Great-Britain and Ireland, and 285,000 packs exported unmanufactured. The value of which, estimated at an average of £.7. per pack, amounts to £. 5,260,724. The quantity manufactured is supposed to amount to the sum of £.12, 434,855, annually, which is circulated amongst industrious artisans. As the whole value of British manufacture, at the above period of calculation, was said not to exceed £.44,350,529, this article alone may be considered as equal in value to one third of all the rest of their produce and manufactures. But what evinces still more the value of sheep to Great-Britain and her dependencies, is, that the wool affords employment to 1,576, 134, out of 4,250,434 people, which are supposed to be the number of the laborious part.

Broad-tailed sheep are found in Tartary, Arabia, Persia, Barbary, Syria and Egypt. Such is the weight of wool on their tails, that Pennant says, some have been known to weigh fifty pounds; to preserve which from

wet

wet, dirt, or other injury, they are ufually fupported by a fmall board running upon wheels.

Of the fheep kind, befide thefe, there are the Strepfiehcros, found in Crete, and other iflands of the Archipelago, the Guinea fheep, and the Moufflon.

The GOAT.

THIS animal differs moft effentially from the fheep, in being covered with hair inftead of wool. Its chief delight is to climb the higheft and fteepeft precipices. They are neither terrified at ftorms, nor incommoded by rain. According to the climate, they will have from two to five kids. Their milk is fweet, nutritive, and medicinal.

The goat is found in every part of the world : every clime feems congenial with its nature: for which reafon it may juftly be called, a citizen of the world. Its age feldom exceeds ten or twelve years.

The Ibex, or Stone-Goat, is faid to have horns two yards long, which increafe by knots annually.

Of the feveral diftinct fpecies of this animal, there are, the Goat of Angora, Syrian Goat, the fmall American Goat, Blue Goat, Juda and Siberian Goat, and the Greenland Goat, the latter of which has horns an ell long.

The

The CAMELOPARD.

THE eamelopard fomewhat refembles the deer in form, without its fymmetry. It has been found eighteen feet high, and ten from the ground to the top of the fhoulder. The hinder parts are fo low, that, when ftanding upright, it greatly refembles a dog fitting. Neither the form nor the temper of this animal adapts him for hoftility or defenee ; he is therefore timorous and innoffenfive, and, notwithftanding its fize, will endeavour to avoid, rather than attaek an enemy. It is chiefly a native of Ethiopia. The extraordinary length of his fore-legs obliges him to divide them when he feeds on vegetables ; to avoid whieh trouble, he fubfifts moftly on the leaves of trees. It is very rare in Europe ; but in earlier times it was known to the Romans, as, among the colleftion of eaftern animals, made on the eelebrated Prænefine pavement, by the direftion of Scylla, the eamelopard is found. It was likewife exhibited by Julius Cæfar, in the Circean games.

It was fuppofed by the Greeks to be generated between a camel and a leopard, from whence it derives its name. It is fo uncommon, that not above one or two have been feen in Europe for many hundred years. Some have neeks fifteen feet long. When they

they walk, they move both their fore-legs together.

The ANTELOPE,

IS principally diftinguifhed from the goat
and deer, by having its horns annulated and
twifted, bunches of hair on the fore-legs, the
lower part being ftreaked with black, red, or
brown, and the infide of the ears having three
white ftreaks.

The Antelope generally inhabits the warmeft climates, thofe of America excepted. It
is equally active and elegant, timid, lively,
and vigilant. Like the hare, its hind-legs
are longeft. It has alfo cloven feet, and per-
manent horns, like the fheep, which are fmaller in the female than the male.

The chafe of thefe animals is a favourite
diverfion in the eaft. In fleetnefs they exceed the greyhound, which frequently caufes the fportfmen to train a falcon to overtake them in the chafe. Their fwiftnefs, has
afforded many beautiful fimilies and allufions
in the eaftern poetry. The eye of the antelope is fuppofed to be the moft beautiful of
any animal in the world, blending brilliancy
with meeknefs. Some of this fpecies form
themfelves into herds of two or three thoufand, and generally feek their food in hilly
countries. Several fyftematic writers have
erroneoufly

erroneoufly ranked this animal among the goat kind; for it forms an intermediate ge- nus between that fpecies and the deer; the texture and permanency of the horns agree- ing with the firft, while their fleetnefs and e- legance accord with the latter.

There is another fpecies of this animal, called the Royal Antelope, or Little Guinea Deer, which is the leaft and moft beautiful of all the cloven-footed race. It is fcarcely nine inches high; and the fmall part of its legs are little thicker than a goofe-quill. It is moft delicately fhaped, refembling that of a ftag in miniature, except that the horns of the male (for the female has none) are hollow and annulated, as in the Gazelle kind. It has broad ears, and two canine teeth in the upper jaw. The colour is as beautiful as the the fymmetry of this little animal, being of a fine gloffy yellow, except the neck and belly, which parts are perfectly white. It is a na- tive of Senegal, and fome parts of Africa. It is fo active that, it will bound over a wall, twelve feet high. It is eafily tamed, when it becomes very entertaining and familiar, but of fo delicate a conftitution that it can bear none but the hotteft climates.

Of antelopes there are, befides thofe before defcribed, the following different fpecies: Common, Blue, Egyptian, Bezoar, Hanaffed, African, Indoftan, White-footed, Swift, Red, Striped, Chinefe, Scythian, Cervine, and Se- negal Antelope.

ANIMALS

ANIMALS of the DEER KIND.

ALTHOUGH the bull and stag do not re-
semble each other in shape and form, yet their
internal structure is very similar. All the
internal difference between them is, that the
deer has no gall-bladder, while the spleen is
proportionally larger, and the kidneys differ-
ently formed.

The first animal of this species that seems
to claim our attention, is the ELK.

The ELK, or FEMALE MOOSE.

THIS animal is a native of both the old
and new continent. In Europe it is called
the elk, and in America the moose deer. It
is sometimes taken in the forests of Germany
and Russia ; but they are found in great numbers in North America. Of the various accounts given of this animal, the following is
esteemed the most authentic.

A female Elk, only twelve months old,
which was in the possession of the late *Mar-
quis of Rockingham*, measured to the top of
the withers fifteen hands ; the length, from
nose to tail, was seven feet ; it had a short
neck, with a thick erect mane, and the body
was

was covered with hoary black hair. It was brought from America, and therefore called a moofe deer. As it was fo young, we may conclude, that, in its wild, and natural ftate, it grows to an amazing height. It is afferted by fome, that in America it grows to the height of twelve feet. This animal is reported to be timorous, gentle, and innoffenfive. It fwims and runs with incredible fwiftnefs. The elk delights in cold countries, where they feed on grafs in fummer, and on the bark of trees in winter. In fnowy weather, they affemble in herds, and feek the firforefts, where they remain, while they can find the leaft fubfiftence from the bark of the trees. At this time they are moftly hunted by the natives of New-England, Nov-Scotia, and Canada, in America; by the inhabitants of Lapland, Norway, Sweden, and Ruffia in Europe; and by the inhabitants of the northeaft parts of Tartary and Siberia, in Afia. The chafe of thefe animals frequently continues two or three days.

The flefh of the elk has an agreeable tafte, and is faid to be nourifhing. The fkin is fo ftrong and thick, as to refift a mufket ball. Its horns are ufed for the fame purpofes as harts horns.

They were formerly ufed in Sweden to draw fledges; but criminals frequentl availing themfelves of their fwiftnefs, to efcape the purfuit of juftice, the ufe of them was prohibited under very fevere penalties.

The

The REIN-DEER.

THIS is the most useful and extraordinary animal of all the deer kind. It is a native of the northern icy regions, and seems adapted by nature to serve that part of mankind who live near the pole. It inhabits further northward than any other hoofed animal; for it is found in Spitzbergen and Greenland. But, in America, it is never seen farther southward than Canada. In Europe, they are also found in Samoidea, Lapland, and Norway. In Asia, they are seen as far as Kamschatka and Siberia. This animal mostly suplies the wants of the Laplanders and Greenlanders; serving them as horses, to draw their sledges over the icy lakes and snowy mountains, which they do with incredible rapidity. Like the cow, they yield all the commodities of milk, cheese, and butter; and as sheep they furnish them with a warm, though homely clothing. The flesh serves them for food, their tendons for bowstrings, and when split, for thread. So that from this quadruped alone, they derive as many advantages as we do from several. The height of a full-grown rein-deer is about four feet six inches. There cannot be stronger proof of the dispensation of divine providence, than in the food which is provided for this animal, when the snowclad face of his country seems to threaten

him

with famine. When not a blade of verdure can be found, on heath, valley, or mountain; trees, bounteoufly affording a black mofs, prove to him a moft ample fuftenance. In the prefervation of this animal, the Laplanders themfelves are much interefted; as, independent of their laborious fervices, the flefh of the rein-deer, is alfo their principal food.

What a contraft do thefe northern countries afford, when compared with thofe of our more clement and fertile climates! The Laplander is obliged to depend on the reindeer for food, clothing, and conveyance, while we have almoft the whole range of nature for our accommodation. Should not this advantage alone excite in us fuch a fenfe of fuperior happinefs, as to render us ever grateful to that Providence, whofe diftinguifhed bounties we enjoy?

The S T A G.

THE colour of this animal is generally of a reddifh brown, with fome black in the face, and a black lift down the hinder part of the neck, and between the fhoulders. The ftag is very delicate in his food; and, during the winter and fpring, feldom drinks. They go about eight months with young, but feldom produce more than one. They breed in May

. D   when

when they carefully conceal their young in
the moſt ſecret thickets. This precaution is
wiſely dictated to them, from their being
expoſed to ſo many formidable enemies, ſuch
as the wolf, dog, eagle, falcon, oſprey,
and all animals of the cat kind. But the
ſtag himſelf is the greateſt enemy to the
young of his ſpecies; infomuch, that the
hind, which is the female of the ſtag, accompanies the faun during the ſummer, to
preſerve it from his depredations. Amongſt
all the enemies of this creature, Man ſeems
to be the greateſt; for in every age, and e-
very country, the human ſpecies have taken
delight in the chaſe of it. Thoſe who firſt
hunted it from neceſſity, continued it afterwards both for health and amuſement. Originally, the beaſts of chaſe were the ſole poſ-
ſeſſors of Great-Britain; they knew no other
conſtraint than the limits of the ocean, nor
acknowledged any particular maſter. But,
when the Saxons eſtabliſhed the heptarchy,
they were reſerved by each ſovereign for
his own particular diverſion. In thoſe unci-
vilized ages, hunting and war were the only
employments of the great; for their active
and uncultivated minds felt no pleaſure but
in rapine or violence.

The other ſpecies of this kind are, the ful-
lo, Virginian, porcine, roebuck, Mexican,
and grey deer.

Stags are ſtill found wild in the highlands
of Scotland, but their ſize is ſmaller than
those

thofe of England. They are likewife to be
feen on the Moors bordering on Cornwall and
Devonfhire ; and on the mountains of Kerry,
in Ireland, where they greatly embellifh the
picturefque, romantic, and magnificent feene-
ry, of the lake of Killarny.

The FEMALE TIBET.

THIS creature, which is the female of
the mufk, gives name to the kingdom of Tibet, a province in China, where it is found,
between the latitude of 45 and 60 degrees.
Thefe animals naturally inhabit the mountains that are covered with pines, delight in
folitude, and avoid mankind : when purfued, they afcend the higheft mountains, which
are inacceffible to men or dogs. It is very
timid, and has fuch a quick fenfe of hearing.
as to difcover an enemy at a very great diftance. The celebrated drug, called *Mufk*,
is produced from the male only, and is found
in a bag about the fize of a hen's egg, on
the belly, which has two fmall crevices
through which it paffes. This drug, when
firft preffed out of the bag, appears like a
brown fat matter ; but it is greatly adulterated by the hunters and dealers, in order to
increafe its weight. Thefe animals are fo
numerous, as to have afforded *Tavernier*
7673 mufk bags, in one journey which he
made,

made, of only three years. Thofe of Muf-
covy are reckoned good, though thofe found
in the kingdom of Tibet are moft valuable.
The Ruffians and Tartars eat the flefh of
the male, notwithftanding its ftrong tafte.
Mufk was formerly in great efteem, as a perfume ; but having been fince found of great
utility in medicine, it is feldom ufed for any
thing elfe. This animal is likewife found in
the Brazils, in India, and in Guinea.

The DROMEDARY.

THIS is the moft temperate of all animals ;
but this difpofition arifes more from neceffity, than from choice or natural moderation.
He is fo admirably formed to crofs the parched deferts, that he will travel eight days
without being thirfty. His hard hoofs are
particularly adapted to travel on the fands of
his native wilds. They are the moft ufeful
beafts of burthen in Arabia, none other being able to bear their loads, or endure the
want of drink fo long ; to enable them to do
which, nature has provided them with a fifth
ftomach, which ferves as a refervoir, from
whence they draw fufficient to quench their
thirft. Camels have been fometimes killed,
in hopes of finding water to flake the parching thirft of the traveller. They are chiefly
                                employed

in affifting the caravans; and as the defarts they crofs afford little more than the coarfeft weeds, they prefer them to the choiceft pafture. He lives forty or fifty years; is about fix feet and a half high, and has callofities on each knee, which greatly eafe him when he kneels down to depofit, or take up his load. A large camel will carry 12 cwt.

The difference between a camel and a dromedary is, that the former has two bunches on his back, the latter only one. There are alfo the Arabian camel, and the Llama camel of America. Camel-hair is imported in great quantities for the ufe of painting.

ANIMALS

IN this kind, animals feem to unite in thofe differences which feparate others.——They refemble the horfe kind in their long heads, fingle ftomachs, and the number of their teeth, which are forty-four. Their cloven feet, and the pofition of the inteftines, are fimilar to thofe of the cow kind. And, in their carnivorous appetite, numerous progeny, and chewing the eud, they refemble the clawfooted kind.

The H O G.

THE hog, in his nature, blends the rapacious with the peaceful kind ; for, though he is furnifhed with arms fufficient to terrify moft, as well as to put the braveft to defiance, he is inoffenfive to all.

He is the moft impure of all quadrupeds; has a moft infatiate appetite, and is of a very fluggifh difpofition. He may be compared to a mifer, who, while living is ufelefs and rapacious, but when dead is confidered a public benefit, by difufing thofe riches he had not fpirit to enjoy when living. The brutality of the hog is fuch, that they frequently devour their own offspring; and, contrary to all other domefticated animals,

when

when impelled by hunger, they will e-
ven devour infants. It is faid to be more
perfect in the internal formation than
any other domeftic animal. The thicknefs of his hide, and the coarfenefs of his
hair, render him infcnfible to blows. He
is naturally ftupid, drowfy, and inactive ;
and, if undifturbed, will fpend half his time
in fleep, from which ftate he never roufes
himfelf but to gratify his voracious appetite,
which if fufficiently fated with food, would
caufe his body to become too heavy for his
legs to fupport ; it would ftill, however,
continue feeding, either kneeling, or lying.
A very remarkable inftance of the voracious difpofition of this animal, is at this
time to be feen in London, in a Warwick-
fhire hog, which, though but a little more
than three years old, meafures nine feet ten
inches in length, five feet ten inches round
the neck, and eight feet five inches in girth.
His weight is ten hundred, two quarters,
and twenty-four pounds. His chief food is
barley-meal and potatoes, which he eats
while lying on his fide ; but what is more remarkable of this furprifing animal is, that he
never drinks.

The hog is reftlefs at every change of
weather, and greatly agitated when the wind
is high. He is fubject to all the difeafes incident to intemperance. When permitted to
extend his thread of life, he will live to
eighteen or twenty years. The fow goes

four

four months, and will often produce fifteen young at a litter.

, The tajacu, pecary, or muſkhog, of South-America, has no tail ; the navel is on its back ; when wounded, it will call its tribe, which are never ſatisfied but in the deſtruction of their antagoniſts or themſelves.

Of the hog, there are, the Guinea, Chineſe, Ethiopian, Indian, hog-rabit, and hog-cow.

The RHINOCEROS.

THIS extraordinary creature inhabits Bengal, Siam, Cochin-China, Quangſi, the iſlands of Java and Sumatra, Congo, Angola, Ethiopia, and the country as low as the Cape. It is next to the elephant in ſize and ſtrength, and has a horn growing on his noſe two feet long. It being our firſt pride to dedicate to truth, we purpoſely omit many fabulous accounts of this animal. Unleſs offended, they are very harmleſs. The fleſh is ſaid to be wholeſome. From its having only one horn, though ſome have been found in Africa with two, this beaſt muſt certainly be the unicorn of holy writ, and the ancients. The ſkin is impenetrable to a muſket ball. Being ſlow and unwieldy in its motions, nature has provided him with a horn, ſo ſtrong, ſolid, and pointed, as to enable

enable him to inflict the moft deadly wounds. Many medicinal virtues are alfo afcribed to this horn, of which cups are frequently made.

His fcent is moft exquifite. He runs in a direct line, his fight not permitting him to fee any thing placed in an oblique direction. Tobacco is his favourite food. The horn was formerly ufed by princes as a cup, in order to detect what poifon might be prefented to them ; for, when any deadly drug is poured on it, it is afferted that it will immediately break into pieces. There is, alfo, another power attributed to this horn, which is, that wine, poured into cups made of it, will rife, boil, and ferment.

This animal was known to the Romans in the moft early ages, and was among thofe of the Præneftine pavement. Ariftotle, who afferts it to have but one horn, calls it the oxyx, and the Indian afs. Auguftus introduced a rhinoceros in an exhibition, which he made on account of his victory over Cleopatra. See Mythology and Roman Hiftory.

The HIPPOPOTAME, *or* SEA-HORSE.

THE hippopotame is as large and formidable as the rhinoceros. The male has been found feventeen feet in length, fifteen feet in circumference, and feven feet in height : the

the legs are three feet long, and the head nearly four. Haffelquift fays, the hide alone is a load for a camel. Its Jaws extend about two feet, with four cutting teeth in each, which are twelve inches in length. The teeth of the fea-horfe are in great eftimation among miniature painters, on account of their never lofing their primitive whitenefs; a quality which the tooth of an elephant does not poffefs. The fkin is fo thick as to refift the edge of a fword or fabre. Contrary to all other amphibious animals, its feet are not webbed. In figure it is between the ox and hog; and is found near lakes and rivers, from the Niger to the Cape of Good Hope, in Africa.

This animal purfues its prey with great rapidity in the water, under which it will remain thirty or forty minutes. They do great injury to the African plantations. Dampier fays they are fo ftrong, that he has feen one overturn a bout with fix men in it : notwithftanding which, they are inoffenfive to all except their natural prey. A convincing proof that Providence has formed the ftrongeft animals to be the moft harmlefs! They never leave the mouth of the frefh-water rivers. The female brings forth her young, which is a fingle offspring, on land. They are taken in pit-falls, and have been often tamed. Their flefh, which is as delicate as veal, is fold like other meat, in the public market.

This

This animal is the Behemoth of Job. It was known to the Romans, and introduced by Auguſtus among other foreign animals that graced his triumph over Cleopatra.

It was worſhipped by the Egyptians, at the city of Papremis, as a ſuperſtitious caution of avoiding any affront to this animal, which they feared might be the caſe, if they refuſed him that deification with which they had honoured ſo many other ſavage beaſts.

The ELEPHANT.

THE elephant is reckoned the largeſt of all land animals, and, next to man, the moſt ſagacious. They grow from ſeven to fifteen feet in height ; and, notwithſtanding their unwieldy bulk, they will ſwim. The trunk with which nature has provided them, and which anſwers the purpoſe of hands to feed themſelves, is formed of many rings. The eyes are extremely ſmall, the legs very ſhort, and the tail like that of a hog. The feet though undivided, have five hoofs round their margins. In the upper jaw are two vaſt tuſks, of ſix or ſeven feet long, from which we obtain our ivory. In droves nothing is more formidable ; wherever they march the foreſt falls before them. When they are thus united, or enraged, it would require an army to repel them ; during their
rutting

rutting time, they are always feized with a tempory madnefs. They cannot live far from water.

The elephant is fo fond of mufic, that he may be learnt to beat time, move in meafure, and join his voice in concert with the inftruments. In Africa it ftill retains its natural liberty. No animal, when tamed, is more courteous, obedient, and affectionate. It kneels to receive its rider. They will draw carriages and fhipping; and frequently carry cannon, and fmall towers, with foldiers in them, to battle, with great courage and perfeverance. They fleep ftanding.— Many have been known to live from 120 to 130 years. The Africans, who take them in pit-falls, very often eat their flefh. A flight wound behind the ear proves fatal to them.

The following is a remarkable inftance of its fenfe, and love of glory : an elephant, being directed to force a large veffel into the water, was found too weak ; on which the mafter, farcaftically defired the keeper to take away the lazy beaft, and bring another. The poor animal was fo affected at the reflection, that he inftantly repeated his efforts, fractured his fkull, and expired.

Let not man boaft of *his* attachment to glory, fince he is thus equalled by the brute creation, in the moft eminent examples.

ANIMALS

ANIMALS of the MONKEY KIND.

THE ape, or monkey clafs, is diftinguifhed from all others by their fimilitude to
man. They have hands, inftead of paws;
their eye-lids, lips, and breafts, greatly refemble thofe of the human race; while their
internal ftructure bears the like conforma-
tion. We recommend, therefore, to thofe
who make their perfons the principal object
of attention, to confider their affinity to this
part of the brute creation, to induce them
to cultivate thofe mental qualifications, which
can alone diftinguifh them from the inferior
claffes of beings!

In the well known ftory of *Peter the wild
boy*, we fee the importance of the cultivation of our infant faculties. This boy was
found by George I. in the woods of Germany, and brought to England in the year
1700, when he was fuppofed to be about ten
or twelve years old; at which time his agility in climbing trees, is faid to have been furprifing. He muft have been loft, or left in the
woods in his early childhood, perhaps foon after he was able to walk; however it might
have happened, his infant impreffions of fociety were loft, and his fubfequent fenti-
ments, being dictated by his favage fituation,
having no opportunity of learning and prac-
tifing fpeech, he continued till his death a

E   mere

mere *ourang outang*.   He could break or
cleave wood, draw water, or threſh in a
barn ; but his rude, narrow mind could
never be enlarged, principally owing to
his not being able to acquire the power
of ſpeech.   This is ſufficient to ſhew
what *we* ſhould be, were we left to ourſelves, and what we owe to the experience
of former ages, for inſtilling into us a proper education, as our faculties expand to
maturity.

The monkey tribe are lively, active, full
of chatter, frolic, and grimace.   Indeed their
actions, as well as their form, ſeem deſign-
ed by nature, to burleſque the ignorant
part of our ſpecies.   In general they are
fierce, untamable, dirty, and diſhoneſt.—
Their greateſt pleaſure is to be perpetually
ſtealing, and hiding their thefts.   Woods
and trees are their chief habitations, where
they feed on fruit, leaves, and inſects.   Such
is their activity, that they will leap from tree
to tree, even when loaded with young.   Being a ſociable animal, they go in companies
or tribes, for the different ſpecies never mix
with each other.   Serpents will purſue them
to the tops of trees, where they frequently
devour them whole.

Although they are not carnivorous, they
will, to gratify their propenſity to miſchief,
rob birds-neſts, both of their eggs and young.
In countries where apes abound, the feathered tribe diſplay great ſagacity in building
                                    their

their nefts as far as poffible beyond their reach.

As thefe creatures differ too much in their fpecies for a general defcription to afford an adequate idea of their nature, we fhall particularly notice the following.

OURANG OUTANG, or *Wild Man of the woods.*

THIS name is given to various animals that walk upright, but which have different proportions, and come from different countries. The ourang outang greatly refembles in countenance, a toothlefs old woman, and approaches nearer to the human race, than any other animal whatever. This creature, indeed, correfponds fo nearly in form to man, that many have expected to find the fame correfpondence. But the contrary being found, difproves that fceptical affertion, that matter forms the nature of the mind. It proves, likewife, that the moft curioufly conftructed bodies are formed in vain, unlefs a correfponding foul is infufed, to direct and controul its operations.

Dr. Tyfon gives the following defcription of one of thefe animals brought from Angola, in Africa.

" The body was covered with black hair,
" which greatly refembles human hair; and
" it

" it was longeſt in the ſame parts, as in the
" human ſpecies. The face was like the
" human face, except the forehead being
" larger, and the head rounder. The jaws
" were not ſo prominent as in monkies, but
" flat like thoſe of a man. The ears,
" teeth, and, in a word, the whole of this
" creature, at firſt view, preſented a human
" figure. And, as he ſo nearly approached
" man in his figure, his diſpoſition was ex-
" ceedingly fond, more gentle, and harm-
" leſs, than the monkey race are found in
" general. Thoſe who were familiar with
" him in the ſhip, he would moſt tenderly
" embrace, open their boſoms, and claſp
" his hands about them.   And, although
" there were other monkies on board, he
" never aſſociated with them ; as if he conſi-
" dered them, as indeed they are, claſſes of
" beings much inferior to him in the ſcale
" of creation. Being accuſtomed to clothes,
" he grew ſo fond of them as to endeavour to
" dreſs and undreſs himſelf. Such parts
" as he could not put on, he took to ſome
" of the company on board, to have their
" aſſiſtance. Like any human creature, he
" would go to bed, place his head on the
" pillow, and cover himſelf with the
" clothes."

One of theſe animals was ſhown in London, in 1738, that would reach himſelf a
chair, and drink tea, which, if too hot, he
would cool in the ſaucer ; he would, likewiſe,

wife, cry like a child. and be exceedingly
unhappy in the abfence of his keeper.

It inhabits the interior parts of Africa, the
iflands of Sumatra, Borneo, and Java.

The ourang outang is folitary in its nature, and fubfifts chiefly on fruits and nuts.
The larger fort are fo ftrong, as to be capable
of overpowering the ftrongeft man. And, as
nature has placed them among the fierceft of
animals, they are provided with fufficient
courage, cunning, and dexterity, to drive
away even elephants from them. They beat
them with their fifts and pieces of wood,
and will even throw ftones at thofe that
offend them. They fometimes carry away
young negroes, efpecially the females, whom
they have been known to treat with the greateft
tendernefs. Le Broffe afferts, that he knew
a woman of Loango, who had lived three
years among them.

The PIGMY APE.

THIS animal has a flat face, with ears
like thofe of a man. It is as large as a cat,
and has olive-brown hair. It fubfifts chiefly
on fruit, ants, and other infects. In order
to find ants, they affemble in troops, and
turn over every ftone in fearch of them.
Africa is the country where they are moftly
found. In animal exhibitions, the pigmy

ape

ape is not uncommon. Their difpofition is very gentle and tractable. The hair on their head feems to come over the forehead like the cowl of a monk. Its hands are remarkably fimilar to thofe of human nature. Of all the various fpecies, this, being the moft harmlefs, is moft fought after by thofe who are fond of making fuch creatures the object of their attention and amufement.

The long-armed ape, called by Mr. Buffon, the gibbun, is a moft extraordinary animal. It walks erect, has no tail, and has fuch long arms, that when he ftands upright he can touch the ground with his hands.

The tufted ape, has a head fo long, that it meafures fourteen inches. It has a long upright tuft of hair on the top of the head, and another under the chin.

There are alfo, the maggot, or Barbary ape ; and the Simia Porcaria.

The BABOON.

THIS animal is about three feet and a half high, has a thick body, ftrong limbs, and long canine teeth. The tail is thick, crooked, and feven inches long. It has a pouch in each cheek, where it depofits its provifions ; which fhows that it is adapted to live in countries where it is liable to meet with a temporary fcarcity ; nature never beftowing
any

any particularity on a being, but in conform-
ity with the neceffity of rendering it capable of living wherever it is placed. Thus
arifes the great difference in animated nature,
from the variety of climates, and not, as fome
have falfely and unphilofophically imagined,
to diftinguifh every part of the creation from
each other.

The baboon fometimes walks erect. Inftead of nails, the hands and feet are armed
with claws, to adapt it for climbing, and render it formidable to thofe natural enemies it
meets with, where it is obliged to feek its
fubfiftence. Forbin relates, that in Siam,
when the men are at harveft-work, whole
troops of them will attack a village, where
the women are obliged to defend themfelves
with clubs, and other weapons, from their
brutal infults. Whatever they undertake,
they execute with furprizing fkill and regularity. When they attack an orchard, they
do it with all the fkill and precaution of an
army in a fiege. They have their fentinels,
and their lines are moft orderly formed. The
female produces but one, which fhe carries in
her arms.

Baboons are not carnivorous; they feed
upon fruits, corn, and roots. Their internal
parts have a greater refemblance to thofe of
quadrupeds than mankind.

The mandril, mentioned by Smith, is a
native of the Gold Coaft. It grows four or
five feet high, and more frequently walks e-
rect

rect than on all fours. When difpleafed, it is faid to weep like a child.

The wanderer is a fmall baboon, remarkable for having a long white head of hair, and a large beard of the fame color.

The little baboon, and the pigtail baboon, are all that remain befide of this fpecies.

Of monkies, there are an innumerable quantity ; we have only room, therefore, to name them as follows : dog-faced, lion-tailed, hare-lipped, fpelted, green, white-eyelid, negro, Chinefe, varied, dove, tawny, winking, goat, four-fingered, weeping, orange, horned, antiqua, fox-tailed, great-eared, filky, and little lion.

ANIMALS.

THE dog, next to the elephant, is the moſt intelligent and friendly to man, of all quadrupeds. Its ſeems beyond the power of ill uſage to alienate his affections from human nature. His beauty, ſwiftneſs, vivacity, courage, fidelity, docility, and watchfulneſs, render him moſt endearing to man. When in his domeſtic ſtate, his firſt ambition and greateſt ſatisfaction, is to pleaſe : he is more humble through affection than ſervility : he waits his orders, and moſt implicitly obeys them. Friendly without intereſts, and grateful for the ſlighteſt favours, he ſooner forgets injuries than benefits; his only aim is to ſerve, never to diſpleaſe.

Numbers of dogs are found wild, or rather without maſters, in Cougs, Lower Ethiopia, and towards the Cape of Good Hope. They go in great packs, and attack lions, tigers, and elephants, by all of which they are frequently killed. Although there are wild dogs, now in South-America, yet this animal was unknown to the new continent, before it was carried there from Europe. This ſhews that the brute creation, like the human ſpecies, may degenerate from a ſtate of refined ſociety, to that of a ſavage nature. In their wild ſtate, they breed in holes, like rabbits; when taken young, they

ſo

so attach themselves to mankind, as never to desert their masters, or return to their savage companions.

The dog is the only animal whose fidelity is unshaken, and almost the only one that knows his name, and answers to the domestic call. No other animal complains aloud for the absence, or loss of his master, or finds so readily his way home, after he has been taken to a distant place.

Of all animals, the dog is most liable to change in its form; the different breds are so numerous, that it is impossible for the most minute observer to describe them; food, climate, and education, all tend to cause deviations in size, hair, shape, and colour. The same dog becomes a different animal, if taken to a different climate from that in which he was bred. Nothing, therefore, but their internal structure, distinguishes this species from every other. They may be said to be all, originally, from the same stock : but which of the kinds can claim the immediate descent, is not yet determined.

The different species of this animal, in its domestic state, are, the shepherd's dog, hound, spaniel, grey-hound, Danish dog, mastiff, bull dog, pup dog, Irish grey-hound, terrier, blood-hound, leymmer, tumbler, lap dog, small Danish dog, Harlequin dog, cur dog, shark, Turkish, and lion dogs.

The

The MASTIFF.

THIS very ufeful and inceftimable animal we have chofen, as firft worthy our particular notice, it being the largeft, and of the moft effential fervice to man.

The maftiff poffeffes great fize and ftrength ; has a large head, with hanging lips, and a noble countenance. This creature is fo formidable, that, Caius fays, the Romans reckoned three of them a match for a bear, and four for a lion. Great Britain was fo famous for maftiffs, that the Roman emperors appointed an officer to fuperintend their breed, and fend them at a proper age, to Rome, for the combats in the Amphitheatre. In England, they are ufually kept to guard yards, houfes, and other places.

In order to try the ftrength of this creature, James I. caufed three of them to be loofed on a lion, which was vanquifhed by their ftrength and courage. Two of the dogs were, indeed, difabled in the combat, but the third obliged the lion to feek his fafety by flight. From the fize, ftrength, and courage, of this noble creature, we may prefume, that nature efpecially formed him for the guardianfhip of mankind ; and being the particular growth of this country, we ought to hold ourfelves greatly indebted to Providence, for fo partial and invaluable a bounty which

which is beſtowed upon us for our accommodation.

The G R E Y - H O U N D.

THIS is the ſwifteſt of all dogs, and
purſues a hare by the ſight, not by ſmell.
Nature, having denied it an acute ſcent, has
recompenſed it with extraordinary ſpeed.
Such is his ſtaunchneſs for hunting, that,
while he keeps the game in view, he will
continue running until he expires, or takes
his prey. The head and legs are long, and
the body ſo exceedingly ſlender, that nothing can be more adapted for fleetneſs. The
grey-hound was formerly eſteemed among
the firſt rank of dogs; which appears by the
foreſt laws of king Canute, wherein it is
enacted, that no perſon, under the degree of
a gentleman, ſhould preſume to keep a grey-
hound.

The various kinds of this animal are, the
Spaniſh grey-hound, which is ſleek and ſmall;
and the oriental grey-hound, which is tall and
ſlender, has very pendulous ears, and long
hair on the tail.

The

The POINTER.

THIS dog is moſt excellent in Spain. It is about the ſize of a bull dog, and ſpotted like a ſpaniel. In diſpoſition, it is docile, and capable of being trained for the greateſt aſſiſtance to the ſportſman who delights in ſhooting. It is aſtoniſhing to ſee to what a degree of obedience theſe animals may be brought. Their ſight is equally acute with their ſcent; from which quickneſs of ſight, they are enabled to perceive, at a diſtance, the ſmalleſt ſign from their maſter. When they ſcent their game, they fix themſelves like ſtatues, in the very attitude in which they happen to be at the moment. If one of their fore feet is not on the ground when they ſcent, it remains ſuſpended, leſt, by putting it to the ground, the game might be too ſoon alarmed with the noiſe. In this poſition they remain, until the ſportſman comes near enough, and is prepared to take his ſhot; when he gives the word, and the dog immediately ſprings the game. Its attitude has often been choſen as a picture for the artiſt to delineate.

Of the other animals of the dog kind, there are, the wolf, fox, jackall, Iſatis, and hyæna.

Of theſe, we ſelect the hyæna and wolf, as the moſt ſingular and remarkable.

F

The

The H Y Æ N A.

THE hyæna is nearly as large as a wolf, which it refembles in the head and body. It is more favage and untamable than any other quadruped, and is continually in a ftate of rage and rapacity; unlefs when feeding, it is always growling. Its gliftening eyes, erect briftles on the back, and teeth always appearing, render its afpect truly terrific. Its horrible howl, refembles a human voice in diftrefs,

The hyæna, from its fize, is the moft terrible and ferocious of all other quadrupeds. It defends itfelf againft the lion, is a match for the panther, and frequently overcomes the ounce. This obfcure and folitary animal chiefly inhabits Afiatic Turkey, Syria, Perfia, and Barbary. Caverns of mountains, cliffs of rocks, and fubterraneous dens, are its chief lurking places. The manfions of the dead are fubject to his violations; for, like the jackall, the putrid contents are, to him, the moft dainty food. It preys upon flocks and herds; but when thefe and other animal prey fails, it will eat the roots of plants, and tender roots of palm-trees.

The fuperftitious Arabs, when they kill a hyæna, always bury its head, left it fhould be applied to magical purpofes, as the neck was formerly by the Theffalian forcerefs: but the
unenlightened

unenlightened Arab muſt be excuſed for this weak opinion, when it is conſidered by the moſt refined and learned ancients, that the hyæna had the power of charming the ſhepherds, and, as it were, rivetting them to the place where they ſtood.

Its voice is a hoarſe, diſagreeable combination, of growling, crying, and roaring.

The fabulous relation of Pliny, reſpecting this creature, is almoſt too abſurd to mention : we, however, relate it, juſt to ſhew how much he debaſed the hiſtory of nature with his fanciful impoſitions. He ſays, that hyænas have been known, not only to imitate the human voice, but to call ſome perſon by his name, who, coming out, was immediately devoured by the ſubtle cruetly of this creature.

In Guinea, Ethiopia, and the Cape, there is another ſpecies of this animal, which is called by Pennant, the ſpotted hyæna.

The WOLF.

THIS animal very much reſembles the dog, both externally and internally, having a long head, pointed noſe, ſharp, erect ears, long buſhy tail, long legs, large teeth, and being covered with longiſh hair. It is of a pale brown colour, tinged with yellow ; though in Canada, it is found both black and white.

The

The principal feature which diftinguifhes its vifage from that of the dog is, that its eyes, which are fierce and fiery, flant upwards, in direction with the nofe.

Though fo near in refemblance to the dog, his nature is entirely different, poffeffing all his ill qualities, without preferving any of the good ones. Thefe animals entertain fuch a natural hatred to each other, that they never meet without fighting or retreating. If the wolf proves victorious, he devours his prey; but the dog, more generous, is content with victory.

They are naturally cruel and cowardly; and will fly the prefence of man, unlefs preffed by hunger, when they prowl by night, in vaft droves, deftroying any perfons they meet; and fuch is their predilection for human flefh, that, when they have once tafted it, they ever after attack the fhepherd in preference to his flock.

The wolf, of all beafts, has the moft rapacious appetite for animal flefh, which nature has furnifhed it with various method of gratifying; notwithftanding which, it moft generally dies of hunger; which is eafily accounted for, when we confider its long profcription, together with the reward formerly offered for its head, which obliged it to fly from human habitation, and feek refuge in woods and forefts.

Wolves were fo numerous in Yorkfhire, in the reign of Athelftan, that it was found
neceffary

neceffary to build a retreat at Flixton, to defend paffengers from their ferocity.   In
France, Spain, and Italy, they are ftill greatly infefted with this animal.   They are alfo
to be found in Afia, Africa, and America ;
but not fo high as the Arctic Circle.

The female goes about fourteen weeks
with young, and brings from five to nine at
a litter.

ANIMALS.

THIS clafs is particularly diftinguifhed by their fharp claws; which they can extend or conceal, at pleafure. They lead a folita-ry, ravenous life; for moft of them not only feek their food alone, but, excepting certain feafons, are enemies to each other. The dog, wolf, and bear, will fometimes live on vegetables ; but the lion, tiger, leo-pard, and all of the cat kind, feed only up-on flefh.

Thefe animals are, in general, fierce, cru-el, fubtle, and rapacious : it is probable, however, that the moft ferocious may be rendered domeftic. Lions have drawn the chariots of conquerors, and tigers have tend-ed thofe herds, which they now deftroy. All animals of the cat kind, though they dif-fer in fize and colour, are allied to each o-ther, in artifice, ferocity, and rapacity.— To fee one, is to know them all. Human affiduity can effect many changes in other creatures ; but, in this kind, all attempts to altar their immutable nature, prove abortive. The dog, cow, and fheep, vary according to their country, but the lion and tiger are the fame, in whatever clime they are found.

This clafs of animals is remarkable for having round heads, fhort nofes, and long

whifkers

whifkers on the upper lip ; they have alfo
thirty very formidable teeth, which are not,
however, fo well adapted for chewing their
food, as for tearing their prey : this fhows,
that nature has formed every creature according to the means they are obliged to a-
dopt to obtain their fubfiftance. Thefe creatures being carnivorous, have teeth particularly adapted to the purpofe ; their claws
are likewife fharp, and ftrong in the gripe,
fo as to enable them to hold their prey, beyond every poffibility of efcape. Not being
capable of running faft, they are formed with
a quicknefs of fcent to difcern their prey,
and feet fo foft, that when they walk, they
may caufe no found which might, prema-
turely, alarm the animal they are going to
furprife.

Although poffeffed of all thefe fierce, and
powerful qualities, they are naturally too
timid to attack any animal poffeffed of more .
ftrength and courage than themfelves. When
they meet with an animal of equal force,
they always retreat, and decline coming to a
conteft..

The LION.

WHAT diftinguifhes this animal's appearance from others, is chiefly his
head, neck, chin, and fhoulders, being covered
with.

with long, shaggy hair, like a mane. It has very strong limbs, and a long tail, with a tuft of hair at the end. The colour is tawny, except on the belly, where it inclines to white. The length of the largest lion, from the nose to the tail, is about eight feet. The lioness is less, and has no mane.

Climate little affects this noble animal. He subsists as well under the frigid pole, as beneath the torrid zone; while most other animals are adapted to live only in particular latitudes.

The lion abounds chiefly in the torrid zone, where they are the largest, and most tremendous. The burning sun, and arid soil, seem to inflame their nature to the greatest height of savage ferocity. In the colder regions, such as Mount Atlas, they are much inferior, both in size, strength, and spirit. The torrid zone, affording but few rivers or fountains, causes him to live in a perpetual fever, which excites a sort of madness, fatal to every animal he meets. It is happy therefore, that this ferocious creature, as travellers in general relate, are daily declining in number. But, perhaps, were they to be entirely extirpated, other animals, on which they prey, might grow too numerous for the safety and welfare of the inhabitants of those dreadful countries. We had, therefore, better leave the proportioning the number of this animal to Him, who measures

all

all things by the fcale of his unerring wifdom and providence.

The eyes of a lion are always bright and fiery, even in death. The paws, teeth, and tongue, perfectly refemble thofe of a cat ; and, in their internal parts, there is fcarcely any difference.

His anger being noble, his courage mag-nanimous, his difpofition grateful, and his conquefts univerfal over all other animals, he is juftly called the king of beafts.

When hungry, he will attack any thing that comes in his way. His teeth are fo ftrong, that he breaks the bones of the ftrongeft animals, which he fwallows with the flefh. He requires about fifteen pounds of flefh per day, and feldom touches any pu-trid body.

The PANTHER.

THIS beaft has been frequently miftaken for the tiger ; which error arofe from its being nearly of the fame fize, poffeffing the fame difpofition to cruelty, and a general enmity to the animal creation. Its chief dif-ference is in being fpotted, and not ftreaked as the tiger.

The panther is found in Barbary, and all the intermediate countries in Africa, that lie between that and Guinea; and is peculiar to

Africa,

Africa as the tiger is to Afia. Although
hunger impels it to attack every thing that
has life, without diftinction, yet it differs
from the tiger, in preferring, at other times,
the flefh of animals to that of mankind.—
Like the tiger, it feizes its prey by furprife,
and will climb trees in purfuit of monkies,
and other creatures which feek an afylum
there. It always retains its fierce, malevo·
lent afpect, and never ceafes to growl or
murmur.

This animal was well known to the ancients, which may be feen by the number con·
tinually introduced by the Romans in their
public fhows. Scarus exhibited 150 panthers
in one fhow : Pompey the great, 410; and
Auguftus, 420. Notwithftanding which, they
are now fwarming in the fouthern parts of
Guinea.

Of the remaining animals of this kind,
we fhall felect the white bear, and the opof·
fum.

The WHITE or POLAR BEAR.

THIS creature grows to a great fize, and
.Ⲧ   is the undifputed mafter of Greenland
and Spitzbergen. When our mariners land
on thofe regions of ice, thefe animals come
down to view them, uncertain whether to
attack or retreat. When fhot at, or wound·
ed,

ed, they endeavour to fly ; but, if they find themfelves incapable, their refiftance never ends but with their death. They live upon feals, carcafes of whales, and fuch human bodies as they can find, or make their prey. Companies of them are fo daring, as to attack crews of armed men, and will even board fmall veffels. From their difpofition to refift all invafion, they feem formed by nature to convince us, that this inhofpitable clime was meant only for their poffeffion, and that it was never defigned by Providence for the abode of the human fpecies. They fwim well, and dive with great agility.— Battles frequently enfue between them and the whales ; in which the latter, from being attacked in their own element, are generally victorious. If, however, they can capture a young whale, they are fufficiently repaid for the danger of meeting the parent.

The affection between the female and their young, is fuch, that they prefer death to parting. The coldeft part of the globe is allotted by nature for the abode of this creature, as they are not to be found further fouth than Newfoundland, unlefs they have been carried involuntarily by floating iflands of ice, on which they had too rafhly ventured in fearch of their prey.

The flefh of this animal is white, and has the tafte of mutton. The fat is melted for train-oil ; and that which is extracted from the feet, is ufed medicinally. The liver is

fo

fo very unwholefome, that it endangered the
lives of three failors who eat fome of it
when boiled.

Dr. Goldfmith relates, that when a Green-
lander and his wife are paddling out at fea,
a white bear will frequently jump into the
boat, and be rowed to fhore like any other
paffenger.

The OPOSSUM.

WHAT diftinguifhes this from all other
animals, and has long excited the wonder of mankind, is a large pouch in the lower
part of the belly of the female, in which the
teats are lodged, and where the young are
fheltered as foon as they are brought forth;
at which time they are blind, naked, finall,
and imperfect. Nature, therefore, has, very providentially, provided them with this
maternal afylum, until they can perfect their
being. But when they are grown ftronger,
they feek fhelter here, as chickens under the
wing of the hen ; here they repofe from fatigue, or feek their food when hungry. On
thefe occafions, the dam moft readily opens
her bag to receive them. The flefh of the
old opoffum is like that of a fucking pig,
the Indian women dye its hair, and weave
it into girdles. The fkin has a very offenfive fmell: the head, which is like that of the

fox, has fifty teeth; the eyes are black, lively, and placed upright; the ears large, broad, and tranfparent; the tail is partly covered with fcales, and partly with hair, which is fuppofed to be that part of the young which cannot be concealed in the pouch, and which nature, therefore, has provided with this armour. The feet refemble hands, having five toes or fingers, with white crooked nails.

The tail of this animal greatly refembles a fnake; by which it will fufpend itfelf on one tree, and, by fwinging its body, throw itfelf among the branches of another. It deftroys poultry, fucking the blood without devouring the flefh: walks extremely flow, and when overtaken, will feign itfelf dead.

It is a native of Virginia, Louifiana, Mexico, Brazil, and Peru.

The other, lefs interefting, animals of the cat kind, are, the domeftic cat, wild cat, ounce, tiger cat, lynx, cougar, fiaguifh, Angora cat, ferval, black bear, brown bear, wolverine or glutton, raccoon, badger, marmoufe, cayopolin, phalanga, and tarfier.

The INDIAN MUSK.

I S a native of Ceylon, of an olive colour, and in length about feventeen inches. Its throat, breaft, and belly, are white, the

fides

ſides and haunches ſpotted, and barred tranſverſely with white. It has large open ears, and a very ſhort tail.

ANIMALS

ANIMALS of the WEASEL KIND.

THIS fpecies is diftinguifhable from other carnivorous animals, by their long and flender bodies, which enable them to creep into very fmall apertures after their prey.. They are called vermin, from refembling the worm in this particular. The form and difpofition of the claws differ from thofe of the cat kind, as they cannot either extend or contract them. - They vary from the dog kind, in being cloathed with fur rather than hair; and differ both in difpofition and appearance. They are cruel, cowardly, and voracious; fubfift moftly by theft; and deftroy all about them before they begin to feed. They fuck the blood of every animal before they eat the flefh.

Of the various individuals of this fpecies, we fhall felect the moft remarkable, beginning with

—————

The CIVET.

THE civet, like the reft of the weafel kind, has a long flender body, fhort legs, and an odorous matter exuding from the glands behind. It is much larger than weafels in general; being in length, from nofe
to

to tail, two feet three inches, the tail fourteen inches, and the body rather thick. It is moftly of an afh colour, fpotted with black; has a long nofe, with whifkers; and eyes that are black and beautiful.

This animal is a native of India, the Philippine Ifles, Guinea, Ethiopia, and Madagafcar. The famous drug, called *Mufk*, is produced from them. To procure which, thofe who keep them provide a box for their habitation, and collect the mufk, by fcraping it three times a week. The male, if irritated, will yield moft. When young, they are fed with millet pap, and a little fifh or flefh; but when old, with raw flefh principally. In their wild ftate they prey on fowls.

Although a native of warm climates, it will live in temperate, and even in cold regions, if carefully defended from the weather.——Great numbers are bred in Holland where they afford confiderable emolument to their owners. The mufk of Amfterdam, being lefs adulterated than any other, is moft efteemed.

The BEAVER.

THE beaver is the only quadruped that has a flat broad tail, covered with fcales, which ferves it as a rudder in the water, and

alfo

alſo as a cart to carry materials for its building on land. The hind feet are webbed, but the fore feet are not, from the neceſſity of uſing them as hands. The fore-part, in general, reſembles a quadruped, and the hind part a fiſh. The teeth are formed like a ſaw, with which they cut the wood they uſe in building their huts, and damming the water out of them. The fur, which is of a deep cheſnut brown, is the moſt valuable material uſed in the hat manufactory. Its length, from noſe to tail, is about three feet ; the tail is eleven inches long, and three broad.

In June and July they form their ſocieties, of two and three hundred, which they continue all the reſt of the year. Wherever they meet, they fix their abode, which is always by the ſide of a lake or river. The ſagacity of this animal is truly worthy the conſideration of the naturaliſt and philoſopher, which it is impoſſible to conſider, without the greateſt humiliation to human pride. When we ſee a beaver, with only its feet, teeth, and tail, capable of building a hut, as commodious for itſelf and young, as a cottage can be rendered to a peaſant, even with the aid of reaſon and mechanical tools, what is the boaſted ſuperiority of man ?

If they fix their ſtation by a river ſubject to floods, they build a dam or pier, which croſſes the ſtream, ſo as to form a piece of water ; but if they ſettle near a

lake,

lake, not liable to inundation, they fave themfelves this trouble. To form this dam or pier, they drive ftakes of about five or fix feet in length, wattling each row with twigs, and filling the interftices with clay, That fide next the water is floped, and the other perpendicular. The bottom is from ten to twelve feet thick, gradually diminifh-ing to the top, which is but two or three feet at moft. This dam is generally from eighty to an hundred feet in length. The greatnefs of the work, compared with the fmallnefs of the architect, however aftonifh-ing, is not more wonderful than its firmnefs and folidity.

The houfes are erected near the fhore, in the water collected by the dams. They are either round or oval, and are built on piles. The tops being vaulted, the infide refembles an oven, and the outfide a dome. The walls, which are two feet thick, are made of earth, ftones, and fticks, and plaiftered with all the fkill and excellence of the moft expert mafon. Every houfe has two openings, one into the water, and the other towards the land.— Their height is about eight feet. From two to thirty beavers inhabit each dwelling; and, in each pond, there are from ten to twenty-five houfes. They have each a bed of mofs; and are fuch perfect epicures, that they dai-ly regale on the choiceft plants and fruits which the country affords.

This

This animal affords that celebrated refinous fubftance, called *Caftoreum*, which is mixed moft fuccefsfully in feveral hyfteric and cephalic medicines. An Oil is likewife extracted from it, called *Oil of Caftor*, which, while it remains in its liquid, unctuous ftate, is ufed for the cure of feveral diforders.

The PORCUPINE.

THIS animal is about two feet long, and fifteen inches in height. The body is covered with quills, from ten to fourteen inches long, and very fharp at the points, growing as feathers in birds. The head, belly, and legs are covered with ftrong brittles. Its whifkers are long, and the ears like thofe of a man. When irritated, its quills ftand erect. The eyes are remarkably fmall, being only about a quarter of an inch wide.

Like the hedge-hog, thefe quills are rather for felf-defence than the purpofe of attacking an enemy. The idea, formerly entertained, that it darted its quills, is found to be erroneous ; they only fhed them when they moult ; which, in fome meafure, fhews their alliance to the bird creation, though not deftined for flight, having neither wings nor feathers. The quills, being found a fufficient defence againft the moft formidable animals, fhow how powerful the weakeft

animals.

animals may be rendered, when under the skill and workmanship of infinite wisdom.

A wolf, it is said was once found dead, with some of the quills of the porcupine sticking in his mouth; no doubt but they must have stuck there when hunger induced him to the rash attempt of devouring this self-defended animal.

The porcupine is generally described to be an inoffensive animal, living entirely on fruits, roots, and vegetables; but some naturalists, particularly Dr. Goldsmith, assert, that they prey upon serpents, with which they live in perpetual enmity. Their method of attacking them is said to be, that the porcupine rolls himself on them, wounding them with its quills, until they expire, when they are immediately devoured by the victor.

It is an inhabitant of India, Persia, Palestine, and every part of Africa. Although not originally a native of Europe, it is found wild in Italy; in which place they have smaller crests, and shorter quills, than those of Asia and Africa.

In Rome, it is sold for food in the public markets.

The SLOTH.

THERE are two kinds of this animal; one of which has two claws on each foot,

and

and is without a tail; the other, three claws
on each foot, with a tail; and are both defcribed under the common appellation of
the floth. It is about the fize of a badger,
and has a coarfe fur, refembling dried grafs;
the tail is exceedingly fhort; and the mouth
extends from ear to ear. The feet of this
animal are fo obliquely placed, that the foles
fcarcely ever touch the ground. The conftruction of its limbs is fo fingular, that it
can move only at the rate of about three
yards in an hour. Thus, unlefs impelled
by hunger, it is feldom induced to change
its place.

The floth inhabits many parts on the eaftern fide of South-America. It is the meaneft, and moft ill-formed of animals. Leaves,
and fruits of trees, are its chief food. It is
a ruminating animal, for which purpofe nature has provided it with four ftomachs.

Although it afcends a tree with great difficulty, yet it cannot defcend without forming itfelf into a ball, and dropping from the
branches to the ground, where the fhock
caufes it to remain for a confiderable time in
a perfect ftate of inactivity. To travel from
one tree to another, at the diftance of one
hundred yards, is, for this animal, a week's
journey.

Every effort, which the floth makes to
move, appears fo painful and difficult, as to
caufe it to utter the moft pitiful cry; which
is likewife wifely given it for its protection;

for,

for being defenceless, as well as incapable
of flight, it could never escape destruction,
was it not that their cry is so hideous, and lamentable in its tone, as to cause every beast
to avoid the sound. How ought we to
admire the wisdom and providence of the
Almighty, who, by the breath only of this
defenceless animal, has raised a bulwark for
its protection!

We should do injustice to the great Creator of the Universe, who never created any
thing in vain, could we suppose any animal
was ever so formed, as to be incapable of
comfort; although the sloth carries every
appearance of misery in its nature, there cannot be a doubt but it has satisfactions peculiarly suited to its station.

The ARMADILLO.

NATURE seems to have reserved all the
wonders of her power for those remote
countries, where man is most savage, and
quadrupeds the most various. She seems to
become more wonderful, in proportion, the
further she retires from human inspection.
But this, in reality, only arises from the attempts of man to rid the country of such
strange productions, in proportion as he becomes more civilized.

The

The armadillo, which is covered with fhells, at the firſt view, appears a round miſhapen ſmaſs, with a long head and ſhort tail. Its ſize is from one to three feet in length. Theſe ſhells which reſemble a bony ſubſtance, cover the head, neck, ſides, rump, and tail. This natural defenſive covering, being jointed, the creature has the power of moving beneath its armour, which reſembles a coat of mail.

As theſe ſhells are only ſufficient to defend the armadillo from a feeble enemy, and not equal to the reſiſtance of a powerful antagoniſt, nature has furniſhed it with a method of encloſing its body within the covert of this armour. Thus, like the hedgehog and porcupine, it is ſecured from danger, without having recourſe to flight or reſiſtance, and becomes invulnerable while in the midſt of danger.

The HARE.

THIS timid and defenceleſs animal is another inſtance of the bountiful care of Providence towards mankind. The hare not only ſupplies us with a delicacy for our table, and a covering for our hands, (the fur being manufactured into hats) but alſo affords us one of the moſt wholeſome of our rural diverſions.

it .

It is an inhabitant of moſt parts of Europe, Aſia, Egypt, Barbary, Japan, Ceylon, and North-America ; but thoſe of Barbary, Spain, and Italy, are much ſmaller than ours. In Wales and France they are generally larger, though not ſo fine a flavour.

This ſolitary animal has, independent of man, a hoſt of enemies, both in the animal and feathered tribes. The fox, polecat, ſtote, and weaſel, hunt them with ſuch unremiting perſeverance, that, notwithſtanding their ſwiftneſs, it is with great difficulty they eſcape their rapacious purſuit. The weaſel will frequently faſten upon the neck of a hare, while on her form, and hold there till it is quite dead, ſucking its blood while running. The kite, hawk, owl, and many other birds of prey, are very deſtructive to young leverets. This perſecuted animal, however, like the rabbit, is ſo prolific, as to afford a plentiful ſupply to thoſe who protect it againſt the unlawful and deſtructive ſnares of the poacher.

The female goes thirty days with young, and brings forth from two to four at a time, with their eyes open ; ſhe breeds ſix or ſeven times a year, and ſuckles her young for twenty days, when her maternal cares ceaſe. After this time they feed on graſs, roots. leaves, corn, plants, and the bark of young trees, to which they are often very deſtructive in nurſeries and plantations. They breed when but a few months old.

Though

Though the hare is reckoned the moſt timorous of all animals in its wild ſtate, it
will, if taken when young, become ſo tame
and familiar, as to ſleep with the grey-hound,
terrier, or pointer ; of which the writer of
this article has been an eye-witneſs. This
ſolitary animal, although not poſſeſſed of the
the wily ſubtilty of the fox, diſcovers a moſt
wonderful inſtinct, which has been given it
for its preſervation. The various ſtratagems
and doubles it makes, when hunted, to avoid
death, would excite the ſurpriſe of every beholder ; nor does it diſplay leſs ſagacity and
cunning, in preventing the poacher from trac-
ing it through the ſnow, by taking the moſt
extraordinary leaps, to elude danger, before
ſhe takes her form.

The RABBIT and the MOLE.

THE great ſimilarity between the rabbit
and the hare, leaves but little to be ſaid by
the natural hiſtorian, or the moraliſt, in its
deſcription. Their figure, food, and natural properties, are nearly the ſame. The
hare ſeeks its ſafety by flight, while the rabbit runs to its ſubterraneous burrow, which
nature has taught her to make with an ingenuity, not to be excelled by the moſt experienced miner. The fruitfulneſs of the rabbit ſo far exceeds that of the hare, that according

H

cording to Pliny and Strabo, they were fo
great a nuifance in the Balearic Iflands, in
the reign of Auguftus, they were under the
neceffity of imploring the affiftance of a mi-
litary force from the Romans to extirpate
them.   A Spanifh hiftorian alfo fays, that, on
the difcovery of a fmall ifland, which they
named Puerto Santo, or Holy Haven, where
they were faved from fhipwreck, they put a
pair of rabbits on fhore, which increafed fo
much in the courfe of a few years, that they
drove away the inhabitants, by deftroying
their corn and plants, who left them to enjoy
the ifland without oppofition.

The MOLE.

AS if nature had meant that no part of
the earth fhould be untenanted, fo the mole is
formed in fuch a manner, as to live entirely
underground.   The fize of this animal is between that of the rat and the moufe, but
without any refemblance of either, being
quite different from any other of the four-
footed race.   It has a nofe like a hog, but
longer in proportion ; inftead of ears, it has
only two holes ; and its eyes are fo remarkably
fmall that it is with the greateft difficulty
they are difcovered.

The moderns, as well as the ancients,
were univerfally of opinion that the mole
                                              was

was totally blind; but Dr. Derham, by the means of a microfcope, difcovered all the parts of the eye known in other animals.

A very fmall degree of vifion being fuffi-cient for a creature deftined to a fubterrane-ous abode, Providence has wifely formed them in this manner: for had they been larger, they would have been continually li-able to injury, by the earth falling into them; to prevent which inconvenience, they are likewife covered with fur. Another wonderful contrivance, to be obferved in nature's works, is, that this animal is fur-nifhed with a certain mufcle, by which it can exert or draw back the eye, as neceffity re-quires.

As a recompence for this defect in the op-tic nerves, the mole enjoys two other fenfes in the higheft perfection; viz. hearing and fmelling; the firft of which gives it the moft early notice of danger, while the latter, al-though in the midft of darknefs directs it to its food. The wants of a fubterraneous a-nimal being but few, fo thofe of the mole are eafily fupplied; worms and infects, in-habiting their regions, being their only food.

Although the mole is generally black, yet it is fometimes fpotted, and has alfo been found quite white. The fur is fhort, and clofe fet, and fmoother than the fineft vel-vet. The length, including the tail, which is about an inch, is feven inches. It breeds

in

in the fpring, and generally brings forth four
or five at a time.

━━━━━━

The JERBOA.

THIS fingular, and, we may fay, pretty
little animal, is a native of Egypt, Barbary,
Paleftine, and the deferts between Balfora,
and Aleppo. It is about the fize of a large
rat ; has dark and full eyes, long whifkers,
broad erect ears, and a head like a rabbit.
The tail is about ten inches long, at the end
of which is a tuft of black hair, tipped with
white. The breaft and belly are of a whitifh colour ; but all the other part of the body is afh-colour at the bottom, and tawny at
the ends. The fore legs are not above an
inch in length, with five toes on each, which
are all furnifhed with fharp claws ; but the
hind legs which are two inches and a
quarter in length, and covered with fhort
hair, exactly refemble thofe of a bird, having but three toes, the middle of which is
the longeft ; they are alfo armed with fharp
claws.

This little animal is as fingular in its motion as in its form ; always walking or ftanding on its hind legs, and ufing the forepaws
as hands, like the fquirrel. It will jump fix
or feven feet from the ground, when purfued, and run fo remarkable fwift that few
quadrupeds

quadrupeds can overtake it. It is a very in-offenfive creature, living entirely on vege-tables. It burrows in the ground, like rab-bits.

In the year 1779, two of them, which were exhibited in London had nearly bur-rowed through the wall of the room in which they were kept.

There is an animal of this fpecies in Sibe-ria, which is a more expert digger than the rabbit itfelf; their burrows are fo numerous in fome places, as to be even dangerous to travellers.

It is related of this latter, that they will cut grafs, and leave it in little heaps to dry; which not only ferves them for food, but alfo makes their habitation warm and comfort-able for themfelves and their young during the winter feafon.

There is alfo the torrid jerboa, fo called by Linnæus from its inhabiting the torrid Zone, which is about the fize of a common moufe; and the Indian jerboa; a fpecimen of which was to be feen in the cabinet of the celebrat-ed Dr. Hunter.

H 2 NATURAL

NATURAL HISTORY.

PART II.

***

BIRDS.

THEIR GENERAL NATURE.

WHILE the forefts, the waters, and even the depths of the earth, have their refpective inhabitants, the air, which includes an immenfe fpace, too elevated for the power of man to explore, is traverfed by innumerable beings, of variegated beauty, called birds; which, in order to facilitate their flight through thofe expanfive regions, with a fwiftnefs to compenfate their want of ftrength, are formed on the following general principles.

Form.—The body of a bird, is made fharp in front, and, when on flight through its native element, it fwells gradually, until the tail is fully expanded, which, with the aid

of

of the wings, ferves it not only as a buoy, but alfo as a rudder to direct its flight.

Plumage.—They are covered with fea-thers, moft admirably adapted to the air they inhabit, being compofed of a quill, containing a confiderable quantity of air, and a fhaft, edged on each fide with a moft volatile fub-ftance, which, with the concavity of the wings, renders the body confiderably lighter than the air ; and thus enables them to explore an immenfe fpace, denied to every other part of the creation.

Sight.—To adapt the fight to the fwiftnefs of their motions, their eyes are not fo con-vex or prominent as in creatures confined to the earth; which not only prevents their being injured by the repulfive force of the air, in their rapid flights, but likewife ren-ders them lefs liable to be touched with the points of thorns, fprays, &c. in their pro-grefs among trees, bufhes, and hedges.— The film, or nictating membrane, with which they occafionally cover their eyes, without clofing the lids, clears and protects them from the glare of fun-beams, as well as from the mifts, fogs, and clouds, with which the air occafionally abounds, when forced to range for food or nefting. The power alfo of extending the optic nerve, gives fuch an acutenefs to their fight, that they can perceive objects more diftinctly, and at a greater diftance, than any other crea-ture.

Hearing.

Hearing.—They have the power of diftinguifhing founds, without any external ear, which would not only impede their flight, but render them liable to many injuries in darting through bufhes, briars, &c.

Smelling.—Their fcent is fo very acute and extenfive, by which they are apprifed of the approach of their natural, as well as artificial enemies, that thofe who decoy ducks are obliged to keep a piece of burning turf in their mouths, to prevent being difcovered.

Internal Structure.—The bones, which are formed fufficiently ftrong to fupport the weight of the body, and the fyftem of its functions, are fo light, as to be fcarcely any additional burthen to the flefh. All their internal ftructure is calculated to increafe the furface beyond the proportion of the folidity of their bodies. In order to render them lighter than the fame portion of air. The lungs and ends of the windpipe branches imbibe air into a number of bladder receptacles. The crop, which is the repofitory for fuperfluous food, fupplies them in long flights, and other times of indifpenfible neceffity. Their food, being generally dry, hard, and crude, they have a gizzard, which, with the help of fand, and other ftony particles they fwallow, aids them in digeftion.

Moulting.—Although birds, from the fimplicity of their ftructure, habitation of the air, and perpetual exercife, are lefs fubject

to

to difeafe than other creatures, yet they are liable to one to which no others are expofed; this is the ficknefs attending the annual renovation of their plumage, which is called their moulting time.

Generation.—In the fpring, when nature affords abundance of food, birds are ftimulated to pair, to increafe their fpecies. Having chofen their mate for the enfuing year, they proceed to thofe official cares which diftinguifh the approach of being made parents. With all the fondnefs of fuch expectations, they proceed to collect materials for their nefts, which they build with the fkill of the moft expert architect. They difcover fo much conftancy to each other, with fuch unabating care and affection in breeding and rearing their young, that they might be taken as examples by the human fpecies.

Habitation.——Birds are particularly attached to the place of their nativity. A rook, if undifturbed, will never quit its native grove; the blackbird and redbreaft are tenacious of their birth-rights; and many others, that are known to emigrate annually from this country, have been found, by frequent experiments, to return to their ufual breeding places.

Migration—Is that paffage of birds from one climate to another, according as they are impelled by fear, hunger, or change of feafons. Many have been the conjectures of
naturalifts

naturalifts and travellers refpecting this extraordinary conduct in particular birds.—
Some have fuppofed that thofe which were
not ftrong enough to fuftain a flight over expanding oceans, collected themfelves in bodies, and repaired to chafms in rocks, or
fought a temporary tomb beneath the waters,
where they remained, in a ftate of torpidity,
until the revolving feafons fhould recall them
to the exercife of their former functions.
Others have imagined, that they actually
fought climes more congenial to their nature
and fubftance, at a time when cold and fcarcity rendered the country of their fojournment both dangerous and inconvenient. The
times of their departure and return are fo
regular, that, in the courfe of five years,
the average has not exceeded more than a
fingle day. Thofe tribes which have not
fufficient ftrength to crofs the immenfe deferts and vaft oceans, fuch as fwallows, mar-
tins, &c. are fuppofed to find a winter fubfiftence in the fouthern countries of Europe,
where the clemency of the feafon feems, moft
hofpitably, to invite them to partake of their
bounties.

It has been obferved, that fome birds,
which migrate in particular climates, are con
ftantly refident in others. According to He-
rodotus, there is a fpecies of fwallow, that
abides perpetually in Egypt ; which muft un-
doubtedly arife from the equality of the feafons in that part of Africa. This property,
therefore,

therefore, is not peculiar to any particular
fpecies of bird, but rather caufed by the difference of the country and climate in which
they are bred.   In Cayan, Java, and other
warm climates, thofe birds, which uniformly
migrate in the cold regions of Norway,
North-America, and Kamtfchatka, are conftant refidents through every change of feafon.   The manner of their departure is too
curious to pafs unnoticed.   They range themfelves in a column, like an I, or in two lines,
refembling the fides of a wedge.   When
they have taken flight, one particular bird
takes the lead ; after going a certain diftance,
he is relieved by another.   In their progrefs,
feveral particulars occur to excite our wonder, as well as our veneration, at that im-
menfity of wifdom, which has formed them
with fo extraordinary an inftinct.   Who acquainted their young with the time, place,
and neceffity of their departure ? and what
can induce them to change the place of
their nativity for a ftrange country ?   Who
caufes the imprifoned bird to feel its captivi-
ty at the time of emigration ; or who is the
herald, to affemble thefe feathered voyagers
and travellers ?   Who is it that forbids one
to depart before the appointed time ?   Who
forms their charts ; or who fupplies them
with a compafs, to direct them over pathlefs
waftes, and tracklefs oceans ? Or who is it
that guides them to thofe countries, where
they reft and recruit themfelves after their
                                    long

long journies, fo as to be enabled to reach their deftined fojournment ? As thefe queftions can only be referred to the wifdom of the great Creator of the univerfe, we cannot avoid learning from them this leffon of humility at leaft ; that, whatever may be the boaft of human reafon, it vaniſhes, when compared with this wonderful inftinct of the emigrative power in birds.

Claſſes.—According to Linnæus, birds are divided into fix claffes, in the following order :

I. The *Rapacious Kind.*—Which are carnivorous, and live by preying on others, or eating the flefh of dead animals. They are diftinguifhed by the beak, which is ftrong, hooked and notched at the point ; by their fhort mufcular legs, ftrong toes, and crooked talons ; by their ftrength of body, impurity of flefh, nature of food, and ferocious cruelty.

II. The *Pie Kind.*—Which are diftinguifhed by their mifcellaneous food, and their females being fed by the males in breeding time.

III. The *Poultry Kind.*—Which are diftinguifhed by their fat mufcular bodies, and pure white flefh. Strangers to any attachment, unlike other birds, they are promifcuous in the choice of their mates.

IV. The *Sparrow Kind.*—Which moftly compofe the vocal and beautiful. Some live

I on

on feed, others on infects. While rearing, they are remarkable fond and faithful.

V. The *Duck Kind*—Are diftinguifhed by their bills, which ferve them as ftrainers for their food; and by their feet, which being webbed, enable them to fwim in the waters, where they chiefly refide.

VI. The *Crane Kind*—Are diftinguifhed by their long and penetrating bills, which enable them to fearch for food at the bottom of waters, near which they chiefly refide; and by their necks and legs, which are proportionable in length.

Having thus briefly given an account of the different claffes, with their diftinguifhing peculiarities, we fhall begin our defcription with thofe which cannot be ranged fyftematically; fuch as the oftrich, caffowary, condour, dodo, &c. which, being of extraordinary fize, and incapable of flying, are not included in the fix claffes before mentioned.

The OSTRICH.

THIS bird, according to naturalifts, is one of the largeft in the world. The head, which is like that of a duck, rifes to the height of a man on horfeback. The body is like a camel, and has two fhort wings, which, though exceeding ftrong, are not expanfive
enough

enough to buoy it from the furface of the earth ; but with their affiftance, added to the length of its legs, it exceeds in fpeed the fwifteft Arabian. It has legs and thighs like a heron, and each foot has three claws covered with horn, the elaftic ftrength of which greatly facilitate and increafes its flight.

Its eggs are fo large, that they commonly weigh fifteen pounds. That they difregard their future progeny, Kolben denies, having feen them fet on their eggs at the Cape of Good Hope. She, however, deferts them by day ; but like other birds, returns to them at night. The climate at the Cape requiring her brooding heat, it is a natural inftinct; but,. in thofe parts of Africa, nearer the equator, we conceive they do, as reported, leave their eggs to be hatched by the fun, but not without the precaution of covering them with fand, and bringing worms and other provifions for the young, when hatched ; for, in birds, as in other creatures, nature conforms to the foil and climate which they are to inhabit. The fimplicity and ignorance of the oftrich is particularly obfervable, in its only hiding its head to fecure its body from the attack of the hunters.

The amazing power poffeffed by this bird, ' of digefting ftones, iron, and other crude fubftances, evinces the wifdom of the Creator, in giving it the faculty of turning to nutriment thofe things which its barren and native deferts only afford.

The

The oftrich feems to fill one of thofe
voids in nature, between the quadruped and
feathered race, as the bat does another; the
former refembling the camel, in the fame proportion as the bat does the moufe.

To the beauty of its plumage this bird
owes its deftruction. But in return, it tri-
umphs over man; for the feathers which its
death affords the purfuers, attend the hearfe
of man to the grave.

The CASSOWARY.

THIS bird, which is found in the fouthern parts of the Eaft Indies, is about five
feet and a half high. The wings are fo fmall,
as to be fcarcely perceptible. It has a creft
on its head, refembling a helmet, three inches
high. Though every feather of this bird is
adapted for flight, none are entirely deftined
for covering. The extremities of them are
armed with five fharp prickles, the longeft of
which is eleven inches. It is defcribed to
have the head of a warrior, the eye of a lion,
defence of a porcupine, and fleetnefs of a
courfer. But though provided thus formidably, it is perfectly inoffenfive. It neither
walks, runs, hops, jumps, or flies; but kick-
ing up one leg, behind, it bounds forward with
the other, with a velocity not to be equalled
by the fwifteft Arabian.   This

This bird, like the oftrich, is extremely voracious of all things capable of paffing its fwallow. The Dutch affert, that it not only devours glafs, iron, and ftones, but even burning coals, without the leaft fear or inju- ry. From its fcarcity, it is generally fuppo- fed not to be fo prolific as the oftrich ; but this may be more owing to their native place being ufurped by man, than from any defect in its nature ; for, both its natural armour and digeftive power, are convincing proofs that it is deftined for the defert, and not for cultivated plains. So that, like other wild creatures, when they have, in vain, difputed with man the poffeffion of their own territo- ries they may have withdrawn themfelves to fome folitary defert, far from the prying eye of man, and for which they are fo pecu- liarly formed.

The EMU,

WHICH is a bird but little known, is fix feet high, refembling the oftrich in form ; and has been reckoned, by travellers and naturalifts, to be of the fame fpecies. It is the largeft bird yet difcovered on the new continent : but it is chiefly found in Guiana, Brazil, Chili, and the immenfe forefts bor- dering on the mouth of the river Plata. Some affert, that it buries its eggs in the

fand, like the oftrich ; but they may be miftaken, as thofe of the crocodile are buried and
hatched in the fame manner.

The D O D O.

T.H E inactive appearance of this bird,
feems to fill another void in nature between
birds and beafts, which is that between the
floth and a more active individual of the feathered tribe. Its body, which is nearly round,
is very ponderous, and covered with grey feathers.   The legs refemble the pillars of a fixed building, but feem fcarcely ftrong enough
to fupport the body ; the neck is thick and
purfy ; and the head has two wide chaps, that
open beyond the eyes, which are large, black,
and prominent ; the bill, which is extremely long and. thick, is of a bluifh white, and
crooked in oppofite directions, refembling
two pointed fpoons laid on the back of
each other.   It has a ftupid and voracious
appearance, which is greatly increafed by a
bordering of feathers, that grow round the
root of the beak, and have the appearance,
of a cowl or hood.——The dodo is, in fhort,
a moft complete picture of ftupidity and de-
formity.

Like the floth, it is incapable either of de-
fence or flight.   It is a native of the Ifle of
France, where it was firft found · by the
Dutch.

Dutch. It is afferted by fome, that the flefh is naufeous, while others, on the contrary, contend that it is palatable and wholefome. This bird grows to fuch an enormous fize, that three or four of them are fufficient to dine a hundred failors. The dodo, by fome, is thought to be the bird of Nazareth, the defcription of it being exactly fimilar to that bird

This feems to be an entire exception to the general nature of birds, both in appearance, as well as activity. If we except the owls, and birds of that defcription, there are fcarcely any but what are agreeable in form, and alert in motion ; but this, on the other hand, appears formed, not only to difguft the fpectator, but to be almoft an immoveable burlefque of the feathered tribe. Were we allowed to give our opinion of the final caufe of creating fo unfeemly a creature, we fhould fay, it was formed as a foil to the various beauties difcovered in the reft of the bird creation.

The GOLDEN EAGLE.

T H I S bird is about three feet nine inches in length, and eight fpans in breadth. Its bill is ftrong, fharp, and crooked : the eye has four lids, to guard it againft exceffive light, and prevent it from external injuries :

the

the toes are covered with fcales ; and the
claws are exceedingly ftrong and formidable.
It is found in the mountainous parts of Ireland, where its fiercenefs has been obferved
to attack cats, dogs, fheep, &c. As it feldom lays more than two eggs, it is a convincing proof that Providence has wifely pre-
vented too great an increafe of what might
prove offenfive, if not deftructive to the poffeffions of mankind. Some of thefe birds.
have been found in Wales.

The male engages in the maintenance of
the young for the firft three months ; after
which time the female undertakes, and continues in this employment, until they are capable of providing for themfelves. The eagle flies the higheft of all birds, and is
therefore called the bird of Heaven. Bo-
chart fays, that it lives a century, during
which period it is continually increafing.
Such is its thirft after blood, that it never
drinks any other liquid, unlefs when fick.
The fwan is the only bird that dare refift
this king of birds. All others, not even excepting the dragon, tremble at its terrific
cry. Not content with preying on birds,
and the fmaller beafts, it will plunge into
feas, lakes, and rivers, after fifh. His fight
is more acute than that of any other bird.
It carries the young on its back to fecure
them from the fowler. The feathers are
renovated every ten years, which greatly
increafes its vigour, as exprefled in the beautiful

tiful fimile of David: *Thy youth fhall be renewed like that of the eagle.* The eagle that
would not quit the corpfe of Pyrrhus, who
had brought it up from a neftling, is a proof
that this fpecies of bird is capable of attachment and gratitude.

There are fixteen other forts of eagles:
namely, the fun, bald, ring-tailed, and black
eagles; ofprey bird; crowned, common, white,
rough-footed, erne, jean le blanc, Brazilian,
Oroonoko, eagle of Pondicherry, and vultur-
ine eagle.

The CONDOUR of AMERICA.

IT is doubtful which this bird is moft allied to, the eagle or the vulture; its force and
vivacity refembling the former, while the
baldnefs of its head and neek are like the latter. No bird can compare with it for fize,
ftrength, rapacity, and fwiftnefs of flight.
It is, therefore, more formidable than the eagle to birds, beafts, and even to mankind.
The rarity of this pernicious and deftructive
bird, is another inftance of the great care of
Providence in proportioning thefe creatures,
according to their utility, or ferocious propenfity; for, were the condour as prolific, or
common as others of the feathered tribe, it
would fpread univerfal devaftation.

Sir

Sir Hans Sloane says, one was shot by Captain Strong, not far from Mocha, an island in the South Seas, on the coast of Chili, as it was sitting on a cliff by the sea side. The wings, when extended, measured, from each extremity, sixteen feet. One of the feathers, which is now in the British Museum, is two feet four inches in length, one inch and a half in circumference, and weighed three drachms, seventeen grains and a half.

According to Garcilasio de la Vaga, several have been killed by the Spaniards, which in general measured fifteen or sixteen feet from wing to wing. To prevent the too fatal exercise of their fierceness, nature has denied them such talons as the eagle. They have only claws, which are as harmless as those of the hen. Their beaks are, however, strong enough to tear off the hide, and penetrate the bowels of an ox. Two of them will attack and devour a cow or a bull; and it has often happened that boys of ten years of age have fallen a prey to them. The inhabitants of Chili are, therefore in continual dread lest their children should be devoured in their absence. In order to allure them, they expose the form of a child, made of a very gluttonous clay, on which they dart with such rapidity, and penetrate so deeply with their beaks, that they cannot disengage themselves. The Indians assert, that they will seize and bear aloft a deer, or a young calf, as easily as eagles do a hare or a rabbit.

Nature

Nature apprifes every one of its approach, by caufing it to make fo great a noife with its wings, as almoft to occafion deafnefs. The body is as large as that of a fheep, and the flefh as difagreeable as carrion. Thus man lofes no food from the providential fcarcity of this terrific and devouring creature. Forefts, not affording room for its flight, are never infeft-ed with its depredations; they, therefore, dwell moftly in mountains, vifiting the fhores at night, when rain or tempeft drive their finney prey thither for fhelter.

They are chiefly to be found in the deferts of Pachomac, where men feldom venture to travel; thofe wild regions being alone fufici-ent to infpire the mind with a fecret horror, affording no other mufic but the roaring of wild beafts, and the hiffing of ferpents; while the adjacent mountains are rendered equally terrible from the vifits of this deftructive bird.

This bird is thought, by naturalifts, to be the fame as the rock, found in Arabia, the Tarnaffar, in the Eaft Indies, and the large vulture, in Senegal.

The ·

The KING of the VULTURES.

THIS bird differs from the eagle, in its indelicate voracity; preying more upon carrion than live animals; which difpofition feems wifely adopted by providence, as a prevention againft the naufeous and epidemical effects that might otherwife arife from carcafes being left to putrify on the earth. Its preying on the eggs of crocodiles, which lay each of them at leaft two or three hundred,.in the fands, is another difpenfation of divine providence, in order to prevent too great an increafe of thofe voracious and deftructive animals.

The form of this bird is diftinguifhed from the eagle, by the nakednefs of its head and neck, though, not being deftined to prey particularly on living birds, &c. their flight is not equal to that of the eagle, falcon, or hawk. But, being allured by putrefaction, their fenfe of fmelling is proportionably exquifite. Happily for us, it is a ftranger to England, while it is found in Arabia, Egypt, and many parts of Africa and Afia. There is a down under the wings, which in the African markets is frequently fold as a valuable fur.

The vulture is confidered fo ferviceable in Egypt, that, in Grand Cairo, large flocks are permitted to refide, in order to devour the

the carrion of that great city, which would otherwife be liable to frequent peftilence.

It is ferviecable, likewife, in thofe countries where hunters purfue, and deftroy animals merely for their fkins ; as they follow, and devour the bodies before they lie long enough to corrupt the air ; which they do fo greedily and voracioufly, as to be unable to fly. But, when they are attacked, they have a power, of lightening their ftomachs, fo as to effect their efcape.

This bird is fomewhat larger than a turkey cock, and remarkable for the uncommon formation of the fkin covering the head and neck, (which is of an orange colour) being bare. The eyes are furrounded with a fkin of a fcarlet colour, and have a beautiful pearl-coloured iris. Although the king of the vultures ftands confeffedly the moft beautiful of this deformed race, its habits are equally difagreeable with the reft.

The flight and cry of thefe birds, being particularly obferved and attended to by the Roman Augurs, muft have arifen from their confidering, where they were moft inclined to direct their flight, from the previous fenfe they had of an approaching flaughter ; which the Romans always flattered themfelves was to enfue of the enemies they were to engage

K   *The*

The GOLDEN VULTURE.

ALTHOUGH this bird is larger, yet in other respects it resembles the golden eagle. It is four feet and a half in length. The lower part of the neck, breast, and belly, are red : the back is covered with black feathers, the wings and tail, with those of a yellowish brown. Though the various species differ very much in respect to colours and dimensions, yet they are all easily distinguished by their naked heads, and beaks partly straight and partly hooked.

In this class are also to be ranged, the golden, ash-coloured, and brown vultures, natives of Europe ; the spotted and black vultures, of Egypt; the Brazillian, and the bearded vultures.

The FALCON.

THE dignified sport of falconry, which formerly distinguished the recreation of the English nobility, has been long discontinued. A person of rank scarcely ever appeared without a falcon, which, in old paintings, are the criterion of titular distinction. Harold, afterwards king of England, was painted with a falcon on his hand, and a dog under his

arm

arm, when he was going on an important em-
bassy. To wind a horn, and carry a falcon
with grace, were then marks of being
well bred. Learning was left for the study
of children, born in a more humble sphere.

In the reign of James I. Sir Thomas Mon-
son gave one thousand pounds for a cast of
hawks. An unqualified person, taking the
eggs of a hawk, even upon his own ground,
was fined and imprisoned, at the pleasure of
the king. Edward III. made it felony to steal
a hawk.

The generous hawk is distinguished from
the baser race of kites, sparrow-hawks, and
buzzards, by the second feather, which in
this kind is the longest; whereas, in the other
kinds, the fourth feather is the longest. They
also possess natural powers of which the other
race are destitute. They pursue their game
with more swiftness and confidence, and, from
their generosity of temper, they are so attached to their feeders, as to become very tractable.

The hawk or falcon pursues the heron,
kite, and woodlark, by flying perpendicular-
ly upwards, which affords the greatest diver-
sion; while other birds by flying horizontal-
ly, diminish the pleasure of the sportsman,
as well as endanger the loss of his hawk.

The Norweigian breed of hawks were of
such esteem in the reign of king John, that,
in consideration of a present of two of these
                    .   birds

birds, that monarch allowed a friend of Jef-fry Fitzpierre to export one hundred weight of cheefe; a very great privilege in thofe days. We learn further, from Madox's antiquities, that the intereft of Richard I. was obtained by the prefent of one Norway hawk, in favour of John, the fon of Ordgar.

The GYR-FALCON.

THIS fpecies of falcon, which exceeds all others, both in fize and eligance, is nearly as large as an eagle. The bill is hooked and yellow, and the plumage moftly white; the feathers of the back and wings have black fpots, in the fhape of hearts: the thighs are clothed with long feathers, of the pureft white: the legs are yellow, and feathered below the knees. This bird is fometimes found intirely white. It was ufed to fly at the nobleft game, fuch as cranes, herons, &c.

In this fpecies of birds may be claffed, the peregrine falcon, facre, mountain, grey, white, Tunis or Barbary falcons, and

The FALCON GENTLE,

WHICH is know from other falcons by the neck, being furrounded with a light yellow ring.

Many miftakes having been made with re-fpect to the names of this fpecies of bird, we think it neceffary to inform our caders, that they are called according to ther times they are taken, after the following names:— If taken in June, July, or Auguft, they are cal-led,   -   -   -   -   -   - *Gentle*
-      -   Sept. Oct. Nov. Dec. - *Pilgrims*
-      -   Jan. Feb. March - - *Antenere*
and if once moulted, it is called   - *Hagar* from the Hebrew, which fignifies *a ftranger*.

The GOSHAWK.

THIS bird, which is larger than the common buzzard, is longer in form, and more elegant in fhape. The breaft and belly are white, beautifully ftreaked with tranfverfe lines of black and white. This fpecies, as well as that of the fparrow-hawk, are dif-tinguifhed by the name of fhort-winged hawks, from their wings, when clofed, not reaching

to the end of the tail.   This bird was formerly much efteemed, and taught by falco-
ners to purfue cranes, wild geefe, pheafants,
and partridges.

The K I T E.

THE kite differs from all the reft of the
fpecies, by its forked tail, flow, floating motion, and being almoft continually on the wing.
Inftead of ufing the wings when flying, it appears to reft on the bofom of the air.   Pliny
fuppofes the invention of the rudder to be
owing to the notice mankind have taken of
the kite, in ufing its tail to direct its flight.
Every bird of the air being capable of efcaping
the purfuit of the kite, it is obliged to fubfift on accidental carnage, which it devours
like a famifhed favage, without the leaft mer-
cy or moderation.

Hunger often makes them fo defperate, as
to attack broods of chickens, ducklings, &c.

It ufually breeds in large forefts, or woody
mountains.   The hen lays two or three eggs,
which, like thofe of other birds of prey, are
larger at the narrow end than thofe laid
by the other fpecies.   When this bird flies
high, it is faid to portend fine and dry weather.
It has been, tho' erroneoufly, reckoned among
                                                birds

birds of paffage. It is twenty-feven inches in length, five feet in breadth, and in weight about forty-four ounces. This bird poffeffing no peculiarity of plumage, we omit giving an uninterefting detail of its feathers: we fhall, therefore, only obferve, that they fometimes differ in color; fome being entirely tawny, while others are variegated.

The COMMON BUZZARD.

THIS bird, which is remarkably fluggifh and inactive, will frequently remain perched a whole day on the fame bough. Frogs, mice, and infects, are its chief fubfiftence. The reafon for preferring which, feems to arife from natural indolence, they being more eafily obtained than birds, which it will not take the trouble of flying after. It lives in the fummer by robbing nefts, and fucking the eggs. In countenance, it more refembles the owl, than any bird of day. Should the hen buzzard be killed, the cock will hatch, and rear the young. They breed in large woods, and generally build on an old crow's neft. The young accompany their parents for fome time after they can fly, which diftinguifhes them from other birds of prey. They vary confiderably in their plumage; fome having brown breafts and bellies, while others,

others are only marked on the breaft with a white crefcent. They are about two feet long, four feet wide, and thirty-two ounces in weight.

Of this fpecies there are alfo, they lioney, moor, and Turkey buzzard; the hen-harri-err, keftril and hobby.

The SPARROW HAWK.

THERE is a great difference in fize between the male and female of this bird; the latter weighing nearly twice as much as the former. They vary alfo confiderably in their plumage ; though the back, head, co-verts of the wings, and tail, are generally of a blue grey. It makes great devaftation among pigeons and partridges.

The fparrow-hawk was in fuch venera-tion among the Egyptians, that they chofe it as the reprefentative of their God Ofiris, and punifhed with death every perfon who fhould kill one. The Greeks confecrated it to A-pollo. It was alfo made one of the fymbols of Juno, from its fixed and piercing fight, which refembled the jealous obfervance of that Goddefs.

The

The MERLIN,

WHICH is the fmalleft of hawks, and not much larger than a thruſh, has been known to kill quails and partidges, and diſplay ſuch courage as to render itſelf as formidable as birds of ſix-times its magnitude.

The female, like that of all birds of prey, is conſiderably larger than the male. It was known to the ancients by the name of Llamyſden.

The GREATER BUTCHER BIRD.

THIS bird leads a life of perpetual hoſtility. It is about the ſize of a blackbird.— From its carnivorous appetite, it participates of the nature of birds of prey, while from its flender legs, feet and toes, it partakes of the nature of thoſe that live upon grain, inſects, &c.

When this bird has killed its prey, it hangs it upon a thorn, as a butcher hangs up a carcaſe, and pulls it to pieces with his bill. Its uſual food is ſmall birds, which it ſeizes by the throat, and ſtrangles in an inſtant. The old and young ſeek their prey in concert. It is ten inches in length, fourteen inches broad, and three ounces in weight. The

back.

back, and coverts of the wings, are of an afh colour, and the fides of the head are white.

Of this fpecies are alfo to be found, the Red-Backed Butcher-Bird, the Wood-chat, and the Leaft Butcher-Bird; which latter, although not much larger than a titmoufe, is a bird of prey. The head is of a fine grey, and beneath each eye there is a tuft of black feathers.

The O W L.

HAVING defcribed the rapacious birds of day, we proceed to thofe of night, which are equally cruel, and more treacherous. That no link in the chain of nature fhould be incomplete, thefe birds employ the night in devaftation, preventing by this means any chafm in the round of time. They are diftinguifhed from all other birds by their eyes, which are better adapted for purpofes of darknefs than of light. Like tygers and cats which fubfift by their nocturnal watchfulnefs, they are endued with the power of difcerning objects, at a time when we fhould conceive it to be totally dark. The idea, however, that they fee beft in total darknefs, is erroneous; twilight, which is the medium between the glare of day, and the gloom of night, being the time they fee with the greateft perfpicu-
ity.

ty. But the faculty of fight differs greatly in the different fpecies.

The note of the owl is truely hedious; and fuch is the antipathy of the fmall birds to it, that, if one appears by chance in the day-time, they all furround, infult, and beat him. So great, however, is the utility of this bird, that one owl will deftroy, in the fame fpace of time, more mice than fix cats.

The white, or barn-owl, which is the moft domeftic, can fee the fmalleft moufe pcep from its hole; while the brown owl is frequently obferved to have a fight ftrong enough to feek its prey in the day-time. Deftined to appear in the night only, nature feems to have thought it unneceffary to lavifh on them any beauties either of form or plumage, as they would have been loft to a general contemplation.

As a fubject of vigilence, this bird was confecreated to Minerva, and feems to fill that chafin betwen quadrupeds and the feathered race, which is obfervable between cats and birds.

The GREAT HORNED OWL.

WHICH is nearly as large as an eagle, has fome feathers rifing from his head which he can elevate or lower, at pleafure. The back, and coverts of the wings, are varied

with

with deep brown and yellow.   It ufually
breeds in caverns, hollow trees, or ruinated
buildings, making their nefts nearly three feet
in diameter.

The LESSER HORNED OWL.

THE wings of this bird are fo long, that
when clofed they reach beyond their tails.
The feathers of the head, back, and coverts
of the wings are brown, edged with yellow:
the tip of the tail is white.

There is alfo a fmaller kind of horned
owl, which is not much larger than the
thrufh.

Of owls, there are alfo the little owl, which
is remarkable for its elegance; the fcreech-
owl, which has blue eyes, and iron-grey feathers; and the brown owl, which remains
all day in the woods.

Notwithftanding this fpecies of birds differ
fo materially, both in fize and plumage, their
eyes are all adapted for nocturnal vifion, to
enable them to feek their food, which they always do by night.   They have ftrong mufcular bodies; powerful feet and claws, for tearing their prey; and ftomachs properly adapted
for digeftion.

BIRDS

THIS clafs is the moft harmlefs, as well as the moft ferviceable to man. It not only furnifhes the table of the epicure with various dainties, but alfo forms a confiderable addition to the neceflaries of life. The rapacious kind may amufe us in the fports of the field, and the warbling fongfter, with its melodious voice, delight us in the grove; but none can equal the effential fervice, and folid advantages of the domeftic poultry. They are a fource of wealth to the peafantry, who keep them at a fmall expenfe, efpecially at farm houfes, and where they have a range of common; which the prodigious influx of eggs and fowls, continually pouring into the markets of this great and opulent metropolis, daily teftify.

They were originally of foreign origin; but time and the climate has fo inured them to us, that they are now confidered as natives; and by their great increafe, form no inconfiderable part of merchandife.

As the rapacious kind are formed for war, this feems equally defirous of peace. They are naturally indolent and voluptuous; have a ftrong ftomach, ufually called a gizzard, which makes them very voracious; while pent up, even, and feparated from their companions, they ftill enjoy the pleafure of eat-

L   ing

ing, and will grow fat, while many of the wilder fpecies pine away, and refufe even common fuftenance.

It is particularly remarkable of this clafs of birds, that, though naturally fond of fo-ciety, their fenfual appetites are fuch, as to admit of no connubial fidelity, which is fuch a diftinguifhing characteriftic in birds of the rapacious kind, fuch as the eagle, &c. whofe connexions, when once formed, never end but with their lives.

The cock, like the bull, wild and irregu-lar in his appetites, ranges from one hen to another, ftruts about the farm yard, like a Sultan in his feraglio, and confiders every one of his fex as his rival and enemy. Carelefs of his progeny, he leaves to the female all the care of providing for the young ; which fhe performs with the greateft maternal care and tendernefs, till they are capable of provi-ding for themfelves.

The hen, equally devoid of fidelity and at-tachment with the cock, when he meets and engages with a rival, ftands an unconnected fpectator of the conflict, and readily receives the embraces of the conqueror.

The cock, when oppofed to a bird of prey, is timorous and cowardly ; but when in oppo-fition to one of his own fpecies, he is na-turally valient, feldom leaving his antagonift until he is killed or taken from him ; many fhameful inftances of which are too frequently

quently exhibited in different parts of the world.

This clafs includes alfo the turkey, Guinea hen, pheafant, buftard, groufe, partridge, and quail; but, as their feveral propenfities are not fo particularly diftinguifhable as the preceding, we fhall content ourfelves with defcribing them in their proper places.

Moft of the birds of this clafs are remarkable for the whitenefs and purity of their flefh, as well as for their bulk. They have ftrong bills to pick up their food, which principally confifts of grain and worms, and fhort concave wings, which render them flow in flight.

═══════

The C O C K.

OF all birds, the cock feems to have been firft reclaimed from the foreft, to gratify the luxury and amufement of man. This bird, in its domeftic ftate, undergoes many variations. In Japan, there is a fpecies of this fowl, which feems to be covered with hair inftead of feathers. Thefe varieties fhow the length of time they muft have been under the dominion of man; the departure from their original characteriftic arifing from the mixture of breeds, brought from different countries, which have been allowed to corrupt, without improving the ftock.

That

That the cock was originally imported from Perfia, is generally acknowledged. It has been, however, fo long in England, that, amongft the ancient Britons, it was one of the forbidden foods.

From the very great length of time this bird has been refident there, we fhould be apt to doubt whether it was natural to any other country, was it not fometimes to be found in the iflands of the Indian ocean, where it ftill retains its wild and natural liberty.

Ariftophanes calls it the Perfian bird, in order to fhow the country where it is produced.

The cock is a very gallant bird, and will fight with his own fpecies, efpecially for the poffeffion of his hens, with amazing courage and perfeverance.

To the bravery of this bird, even princes themfelves, in different parts of the world, have, to their fhame be it fpoken, owed a principal part of their amufement. Heathens might have fallen into this error; but that a race of people, calling themfelves Chriftians, who are ftiled the patrons of compaffion and humanity, fhould take delight in fetting thefe inoffenfive birds to deftroy each other, can only be attributed to a barbarous propenfity in human nature, which we cannot but lament.

Exclufive of this, there are two other fpecies of cocks, called the Hamburgh and

. Bantam

Bantam cocks; the latter of which is well known, by its diminutive fize and feathered legs.

The PEACOCK.

THE Italians have obferved, not unaptly, that this bird has the plumage of an angel, the voice of a demon, and the appetite of a thief. They were originally from India, and are ftill found in vaft flocks in the iflands of Ceylon and Java. The beauty of the peacock deprived it firft of its liberty; which proves to demonftration, that beauty is not confined to the deftruction of the human fpecies. So early as in the time of Solomon, according to the tenth chapter of the firft book of Kings, apes and peacocks are found among the articles that were imported from Tarfhifh. They were fo much efteemed by the Greeks, that a pair of them was reckoned worth upwards of thirty pounds fterling. When firft introduced into Greece, they were made a public exhibition. Hortenfius, the orator, was the firft who ferved them up as an entertainment for the table. They were afterwards confidered the choiceft of viands, and one of the greateft ornaments of the feaft; but their palatable fame foon declined, as may be obferved by the conduct of Francis I. who ferved

L 2 them

them up in their plumage, by way of ornament, not as a dainty.

To defcribe the peacock as concifely as
poffible; we have only to obferve, that the
head, neck, and beginning of the breaft,
are of a deep fhining blue : on the crown, is
a tuft of green feathers ; and the tail, which
may be faid to vie in fplendour with the
rainbow, (the colours being fo beautifully
intermixed) they difplay with all the feem-
ing vanity, of a conceited beauty. The gold,
chefnut, green, and blue of the eyes, are fo
happily difpofed, that they form the fineft
harmony, and moft beautiful contraft of colour that can poffibly be conceived. The
bird himfelf is fo fenfible of this fuperiority
of plumage, which certainly exceeds every
thing of the kind in nature's works, that he
is never fo proud as when he exhibits this
unrivalled work of the Divine Artift, to
whom he is indebted for his form and exiftence.

The PHEASANT.

THE plumage of this bird is fo beautiful, that many efteem it next in rank to the
peacock. Crœfus, king of Lydia, when
feated on his throne, arrayed in all the fplendor of the Eaft, afked Solon, if he had ever
feen any thing fo fine? To which the philo-
fopher

fopher replied, that, after having feen a pheafant, no other finery could aftonifh him.

Although the pheafant is, certainly, a moft beautiful and elegant bird, yet there are many others, as well as the peacock, which can vie with it in plumage. Its chief beauties are in the eyes, which are yellow, furrounded with fcarlet, and fpotted with black; black feathers, intermingled with a gloffy purple, adorn the fore part of the head; while the top of the head, and the upper part of the neck, are tinged with a darkifh, fhining green : the back, fides, breaft, and fhoulders, are of a black colour, changing to purple, according to the fituation of the fpectator, under which purple is a tranfverfe ftreak of a gold colour.

The tail is about eighteen inches long; the legs, feet, and toes, are of a horn colour; and two of the toes are connected by a membrane.

This bird is not only beautiful to the eye, but extremely delicate to the tafte. But, as if it difdained the commerce of man, it has left him to take fhelter in the woods and forefts; to which unlimited freedom may be attributed the exquifite flavour of its flefh.

. The

The GOLDEN PHEASANT of CHINA.

THIS bird, which is said to excel all the rest in beauty, is so prolific, that, when in its wild state, it will lay twenty eggs, and upwards, being double the number they lay when domesticated. The pheasant, of all wild birds, is most easily shot.

Besides those already mentioned, there are the horned Indian, red China, white China, peacock, and Brazilian pheasants.

The BUSTARD,

IS the largest native land bird of Britain; the male generally weighing twenty-five pounds. It is about nine feet broad, and four feet long. The head and neck are of an ash-colour, and the back is barred transversly with black, bright, and rust colour: the greater quill feathers are black; those on the belly are white; the tail, which consists of twenty feathers, has broad red and black stripes; and the legs are of a dusky hue.

The female is about half the size of the male. They were formerly much more numerous than at present; but the increased cultivation of the country, added to the extreme delicacy of its flesh, has caused a great

decrease

decreafe of the fpecies. Another circumftance, equally unfavourable to this bird, is its ama-zing fize, which renders it fo unwieldy and flow in flight, as to render it almoft impoffible to efcape the aim of the fportfman.

Buftards are principally found on Salifbury Plains, Newmarket and Royfton Heaths, Derfetfhire Uplands, and thofe of Marfh or Lothian, in Scotland. They run very faft; and, although flow in flight, will, when on the wing, continue their progrefs, without refting for feveral miles. It is with fuch dif-ficulty they take flight, that they are fre-quently run down by grey hounds. They feldom wander above twenty or thirty miles from their haunts. They live on berries, which grow on the heaths, and on earth-worms, that are found on the downs before the fun rifes.

As a fecurity againft drought, nature has furnifhed the male with a pouch, that will contain near feven quarts of water, with which, it is fuppofed, they accommodate and fupply the female while fitting, or the young, until they can fly.

It lives about fifteen years, but cannot be domefticated from the want of a fufficient fupply of the food which they delight in, which they can only obtain in their natural ftate.

There are two other fpecies of this bird, which are called the Indian buftard and little buftard.

BIRDS

BIRDS of the PIE KIND.

THIS clafs of birds, though not formed for war, delight in mifchief, and are perpetually harraffing other birds, without the leaft apparent enmity ; and includes all that noify reftlefs, chattering, tribe, from the raven to the woodpecker, which hover about our habitations, and make free with the fruits of our induftry.

Though they contribute the leaft of any birds to the pleafures or neceffities of man, they are as remarkable for inftinct, as for their capacity to receive inftruction ; cunning and archnefs are peculiar to the whole tribe.   They have hoarfe voices, flender bodies, and a facility of flight which baffles the purfuit of all the rapacious kind.   Of this clafs we felect the following, as moft deferving our attention.

The TOUCAN,

WHICH in fize and fhape refembles a Jack-daw, has a remarkable large head, to fupport an cnormous bill, which, from the angles of the mouth to the point, extends fix inches and a half in length, and upwards of two inches in breadth, in the broadeft part, not much thicker than parchment. Some naturalifts tuarlift have thought, but erroneoufly, that the toucan had no noftrils; this miftake, in all probability originated from their being placed in the upper part of the bill, and, confequently, neatly covered with feathers.

Between the white on the breaft, and the black on the belly are a number of red feathers, moft beautifully formed in the fhape of a crefcent, with the horns pointing upwards. The toes are difpofed in the fame manner as thofe of the parrot, two before and two behind.

ꞌ The toucan is fo eafily tamed, that it will hatch and rear its young in houfes. Its chief food is pepper, which it is faid to devour like a glutton. Pozzo, who bred one of thefe birds, fays, that it refembles a magpie both in voice and motion. Naturalifts feem to think, that the toucan ufes its tongue to all thofe purpofes for which other birds ufe their bills. This naturally accounts for the thinnefs of the beak, which feems only calculated as a fheath for the tongue, which is very large and ftrong.

This bird inhabits only the warm climates of South-America, where it is much efteemed for the delicacy of its flefh, and beauty of plumage. The feathers of the breaft are particularly admired among the Indians, who pluck them from this part of the fkin, and, when dry, glue them to their cheeks, which they reckon an irrefiftible addition to female beauty.

When

When we contemplate the bird creation, we cannot confider without amazement, how varioufly nature has formed their bills, wings, feet, and bodies, according to their different wants and peculiarities, occafioned either by fituation or difpofition; a more ftriking inftance of which cannot be adduced than in the bird juft defcribed.

The GREAT SPOTTED WOODPECKER.

THIS bird is about nine inches long, fixteen inches in breadth, and two ounces three quarters in weight. The bill is of a black horn colour, and the forehead pale buff; the crown of the head is of a gloffy black, and the hind part is marked with a deep rich crimfon fpot; the cheeks are white, bounded beneath by a black line, which paffes from each corner of the mouth, and furrounds the hind part of the head; the neck is incircled with black; the throat and breaft are of a yellowifh white; the back, rump, coverts of the tail, and leffer coverts of the wings, are black. The webs of the black quill feathers are elegantly marked with round white fpots. The four middle feathers of the tail are black; the next are tipped with dirty yellow; and the ends of the two outermoft are black. The legs are of a red colour.

The

The colours of the female are the fame as in the male, except the crimfon fpot on the head.

The G R E E N WOODPECKER.

OF this bird there are many kinds and varieties, forming large colonies, in the forefts of almoft every part of the world. The wifdom of Providence, in the admirable formation of creatures according to their refpective natures, cannot be better exemplified than in the birds of this genus.

Woodpeckers, feeding entirely upon infects, and their principal action being neceffarily that of climbing up and down the trunks or branches of trees, have a long flender tongue, armed with a fharp bony end, barbed on each fide, which, with the affiftance of a curious apparatus of mufcles, they dart to a great depth into the clefts of the bark, from whence they draw out the lurking infects.

When this bird difcovers a rotten, hollow tree, it cries aloud, which alarms the infect colony, and puts them into ccnfufion; by which means it is better enabled to get at the prey. By thus deftroying thefe infects, which are found fometimes on trees not entirely decayed, it fhould feem as if nature had formed this bird for the exprefs purpofe

M of

cleanfing fuch trees, as they are generally obferved to thrive and flourifh, after they have left
them.   They are likewife very ufeful in deftroying ants, on which they feed, as well as
on wood-worms and infects.   To take ants,
they adopt the following curious ftratagem:
they dart their red tongues into the ant-hill,
which the ants, from the refemblance, fuppofing to be their ufual prey, fettle upon it
in myraids, which is no fooner done than
they withdraw their tongues, and devour
them.

The green woodpecker is about thirteen
inches long, twenty-one inches in breadth,
and weighs fix-ounces and a half.   The bill
is hard, ftrong, and fhaped like a wedge.   Dr.
Derham fays it has a neat ridge running along
the top, which feems as if it was defigned by
an artift, both for ftrength and beauty.   The
back, neck, and leffer coverts of the wings,
are green, and the rump is of a pale yellow.

To thefe may be added, the leffer fpotted,
and Guinea woodpeckers.

The BIRD OF PARADISE.

ACCORDING to fome naturalifts,
there are nine different forts of this bird;
but Mr. Edwards defcribes only the three
following: viz. The greater bird of Para-
                                        dife,

dife, the king of the birds of Paradife, and the golden bird of Paradife.

The bird of Paradife, as defcribed by More-grave, is about the fize of a fwallow. The feathers about the beak are as foft as filk, green and brown above, and black below ; the upper part of the neck is of a gold colour, but lower down, it is gold mixed with green : the long feathers on the fides, near the rife, are of a gold colour, and the other parts of a whitifh yellow.

The king of the birds of paradife, mention-ed by Clufius is the leaft of the fpecies.

The golden bird of Paradife, has a gold eo-loured neck and beak ; the feet and toes are yellow ; breaft and back pale orange colour ; and the large feathers on the wings and tail, are of a deep orange colour.

The idea that thefe birds have no feet, is proved to be an error by Mr. Ray, who fays, their feet are neither fmall nor weak, but large and long, armed with crooked talons, like birds of prey.

The great beauty and variety difplayed in every part of the creation, continually af-fords, to the contemplative mind, frefh in-ftances of the power, wifdom, and goodnefs, of the Divine and Almighty architect.

The bird of Paradife, which is a native of the Molucca Iflands, exceeds every other bird of the pie kind in beauty ; a proof, that thofe groves which produce the richeft fpices, pro-duce alfo the fineft birds. The inhabitants,

fenfible

fenfible of the fuperiority of thefe birds, call them, by way of pre-eminence God's birds.

They migrate with their king (which is fuperior both in fize and plumage) about Auguft, when the ftormy feafon begins, and return when it is over.

There are two other birds of Paradife ; one of which is found in the Ifland of Ceylon, but has never yet been defcribed ; the other is called the pied bird of Paradife, has a blackifh bill, like a duck, and a tail nearly as long as a magpie.

_____

The C U C K O O.

T H E note of this bird is known to all the world; but its hiftory and nature remain yet undifcovered. Some naturalifts have af-ferted, that it devours its parent, changing its nature with the feafon, when it becomes a fparrow-hawk. But thefe fables are now fufficiently refuted. It however, ftill remains a fecret where it refides, and how it fubfifts in winter.

The claws and bill of the cuckoo are much weaker than thofe of other rapacious birds. It is diftinguifhed from all others, by its note, and the round prominent noftrils on the fur-face of the bill. The head, the upper part of the body, and the wings, are beautiful-ly

ly ftriped with tawny colour and tranfparent black ; the legs are very fhort, clothed with feathers down to the feet ; and
it has a large mouth, the infide of which is
yellowifh.

This bird is the harbinger of fpring at
which time it returns, to glad the hufband-
man with its wonted note, as a fignal that
nature now refumes her vernal beauties. The
note, which is a call to love, is ufed only by
the male, and continues no longer than the
pairing feafon.

The young are generally nurfed by a wa-
ter-wagtail or hedge-fparrow, their parents
always unnaturally deferting them.

The note of the cuckoo is pleafant though
uniform ; and owes its power of pleafing to
that affociation of ideas which frequently
render things agreeable, that would, otherwife, not be fo in themfelves. Were we to
hear the cuckoo on the approach of winter,
we fhould think it a moft lamentable noife ;
but, hearing it as we do, at the approach of
fpring, we cannot avoid thinking it the moft
agreeable, from its being attached to all
thofe enjoyments, with which we know nature is then teeming for our accommodation.

It is about fourteen inches in length, twenty-five in breadth, and weighs five ounces,
little more or lefs.

M 2   BIRDS

BIRDS of the SPARROW KIND,

DESCENDING from the larger to the fmaller kinds, we come to this clafs of birds, which live chiefly in the neighbourhood of man, whom they feem to confider as their beft friend, filling his groves and fields with harmony, that elevates his heart to fhare their raptures. All other birds are either mute or fcreaming ; and it is only this diminutive tribe that have voices equal to their beauty. Great birds feem to dread the vicinity of man, while thefe alone remain in the neighbourhood of cultivation, warbling in hedge-rows, or mixing with the poultry, in the farm-yard.

They are remarkably brave ; often fighting until one of them yields up its life with the victory. When young, they are fed upon worms and infects ; but, when grown up, they feed principally upon grain. As they devour great fwarms of pernicious vermin, which deftroy the root before the vegetable is grown, they are particularly ufeful to the farmer and gardener.

The beft vocal performers of this mufical tribe, are, the nightingale, thrufh, black-bird, lark, redbreaft, blackcap, wren, Canary-bird, linnet, goldfinch, bulfinch brambling, yellowhammer and fifkin.

This

This clafs being too extenfive to be fully defcribed in fo fmall a volume, we fhall felect only a few of the moft curious, beginning with

The BLACKBIRD.

THIS bird, which is the herald that ufhers in the welcome fpring, feems, by its melody, to awaken the reft of the feathered creation from their lethargy, and allure them to the pleafures of the approaching feafon. They generally breed about the latter end of March, or beginning of April, laying four or five eggs, which are of a blueifh green colour, and irregularly marked with dufky fpots. Their nefts are conftructed, in a very ingenious manner, with mofs, twigs, and fibres of roots ftrongly cemented ; the infides being plaiftered with clay, and covered with hair, and other foft materials. They ufually build in hedges, near the ground, and before the foliage expands, which, added to the magnitude of the neft, renders it eafy to be difcovered.

The plumage of the male, when at full age, which is a year, is of a fine deep black, while the bill, as well as the edges of the eyelids, are of a beautiful bright yellow ; but before they attain this age, the bill is dufky, and the plumage of a rufty black.

They

They continue finging till the moulting feafon draws near, when they naturally defift; they will, however, when they have done moulting, refume their note for a fhort time previous to the winter.

The STARLING.

THE ftare, or ftarling, may be diftin-guifhed from the reft of the fparrow tribe, by the variegation of its feathers. which in fome lights fhow a gloffy green, and in others a beautiful purple. The feathers of the head, neck, and upper part of the breaft are black, interfperfed with feathers of different co-lours, which caufes it to vary, as above def-cribed.

Starlings affemble in vaft flocks during winter, and feed upon worms and infects; but, on the approach of fpring, they meet in the fields, as if to confult ; during this time, which laft feveral days, they feem to abftain from all kind of nourifhment.

Such is the capacity of this bird to receive inftruction, that it will imitate the human voice to the greateft nicety. Sterne, in his Sentimental Journey, gives a very entertain-ing account of one of thefe birds which he met with on his travels.

If a ftarling is taken when about ten days old, and properly taught, it is a very valua-

ble

bie bird, and will fetch frequently five or six guineas.

The GREAT TITMOUSE.

T H I S bird, which is alfo called the ox-eye, is about fix inches in length, nine inches broad, and in weight half an ounce. The bill is black, ftraight, and about half an inch long ; the tongue is broad, ending in four filaments ; the head and throat arc black, cheeks white, back, and coverts of the wings, green ; quill-feathers dufky, tipped with blue and white ; the leffer coverts arc blue, the greater ones tipped with white ; the tail, which is about two inches and a half in length, is black, edged with blue.

Although thefe birds occafionally vifit our gardens, yet they chiefly inhabit the woods, where they build their nefts in hollow trees, laying nine or ten eggs. Their food confifts, principally, of infects, which they find in great numbers in the trees. Thus we perceive, that birds are formed, not only to delight the ear and pleafe the eye, but alfo to ferve us, by deftroying thofe vermin, which do incredible mifchief to our ruftic poffeffions. As we can have no enjoyment, however, without fome mixture of alloy, the titmoufe frequently injures our fruit-gardens, by deftroying the tender buds.

Like

Like the woodpecker, it is continually running up and down the trunks of trees, searching for food.

The LARGE-CRESTED HUMMING-BIRD.

THERE is a great variety in this species of birds, which, although the smallest of the feathered tribe, are by far the most beautiful, inoffensive, and delighting. They are from the size of the wren to an humble bee. What a beautiful contrast does this little bird afford, in the scale of creation, when presented by the side of the largest ostrich, forming the two extremes? and how can we sufficiently admire the workmanship of Providence, in having created such varieties for our use, entertainment, and assistance.

In America they swarm like bees, ranging from flower to flower, extracting the sweets; in which they seem to connect the insect and bird creation together.

The head of this beautiful bird is adorned with a crest, green at bottom, and bright gold-colour at top; the body, and under the wings, is brown and green intermixed, and glossed with a beautiful red; the bill is black, straight, and slender: the eyes black and sparkling.

They

They are called humming birds, from the noife produced by the motion of the wings. Their nefts, about half the fize of a hen's egg, are curioufly fufpended at the end of the twigs of an orange or pomegranate tree.

There are alfo, the larger humming bird, long-tailed black-capped humming bird, little humming bird with crooked bill, green, and afh-coloured humming birds.

The HOOPOE.

THIS very handfome feathered vifitant, according to the ingenious Mr. Walcot, in his Synopfis, juft publifhed in quarto, anfwers the following very curious and interefting defcription:

On the top of the head is a creft, confifting of a double row of feathers, the higheft of which are about two inches in length, of a pale orange colour, with black ends; the neck is of a pale reddifh brown, the upper parts of which are croffed with broad bars of black and white; the leffer coverts of the wings are of a light brown, and the lower parts white; the tail, which is white, confifts of ten black feathers, which are marked with a white crefcent; the legs are black. It is twelve inches in length, and nintcen in breadth.

A few

A few of these beautiful birds migrate to this country in the summer, and feed on infects. It is said to make no nest, but to lay out seven ash-coloured eggs, in the holes of trees, walls, or on the ground.

The KING-FISHER.

THIS beautiful bird which inhabits almost every country, may be said to vie in elegance of plumage, with the parrot, the peacock, or even the splendid shadings of the humming bird. It is larger than the swallow; mostly frequents the banks of rivers, and makes its nest at the root of some decayed tree, which it lines with the down of willow. They lay from five to nine white eggs before they sit, and hatch twice a year. In this bird we have an instance of parental and conjugal affection, which might shame many of the human race; as a proof of which that ingenious author, Reaumer, says, that he had a female of this species brought to his house, upwards of three leagues from her nest. After having admired her beautiful colours, he let her fly again, when the fond creature was observed instantly to return to the nest where she had just before been made a captive; when joining her mate, she began to lay again, though it was the third time, and

the

feafon very far advanced. She had feven eggs each time. The fidelity of the male exceeds even that of the turtle. While the hen is fitting, and during the helplefs ftate of her callous brood, he fupplies her with fifh, which he takes with the greateft expertnefs, and in large quantities ; infomuch, that at this feafon, fhe, contrary to moft other birds, is fat, and in fine feather.

Several writers have confounded the halcyon with the king fifher. The halcyon, it is faid, breeds in May, in the banks of ftreams, near the fea ; after the firft hatch is reared, it returns to lay again in the fame neft. Pliny and Ariftotle fay, that the halcyon is common in the feas of Sicily; that it fits only a few days, in the depth of winter, in a neft that fwims on the fea; during which time, it is faid, the mariner may fail with the greateft fafety. But another author, with more probability, fays, that the little halcyon bird is found on the fhores and rocks up the Mediterranean, near Sicily ; that, at the latter end of fummer, fhe builds a neft, with fifh bones and fea weeds, fo curious and impregnable, as to fwim and hatch her young on the fea, which at that time is particularly calm and ferene. This has given rife to a proverbial faying, when we allude to any particular period of our lives, wherein we have experienced uninterrupted happinefs, which are called *halcyon days.*

N   BIRDS

BIRDS *of the* CRANE KIND.

THIS clafs is inferior to every other in building their nefts, being lefs curious than thofe of the fparrow kind ; the method they ufe to obtain their food, is alfo lefs ingenious than thofe of the falcon kind ; the pie kind excel them in cunning ; while the poultry kind are more prolific. None of this kind being, therefore, protected by man, they lead a precarious life in fens and marfhes, where they feed upon fifh and infects ; for which purpofe nature has provided them with long necks, to enable them to dive for their prey, and long legs to keep their bodies dry and clean.

Thofe only which feed on infects are eat-able.

The STORK.

THIS bird is fimilar to the crane, but more remarkable both in figure and difpofi-tion. The feathers are white and brown; and the nails are flat, like thofe of a man. It makes no other noife, but that of clacking its under bill againft the upper. Contrary to the general difpofition of nature, it has as much, if not more filial affection toward

its

its parents, then paternal affection for its offspring ; for, when the old ones are so far
advanced in years, as to be incapable of providing for themselves, the young ones will
serve them with food in the hour of necessi-
ty, cover and cherish them with their wings,
and even carry them on their backs to a great
distance.. What an example is this of filial
piety! Who can observe this affectionate
bird, feeding and defending its aged and helpless
parent, till death relieves them from their
anxiety, without exclaiming, *O ye children
imitate this amiable example* ; *let not a simple bird upbraid and condemn you ;· but, on
the contrary, let it stimulate you to the dif-
charge of this most pleasing duty ; let it recall to your mind the anxious days, and sleep-
less nights they have endured in nursing, pro-
tecting, and promoting your welfare ; and you
will not fail to imitate· the stork, in soothing
their decline of life, with the lenients of your
love, care, obedience and gratitude.*

The HERON.

THIS bird may be distinguished from the
crane and stork, by its smaller size ; by the
bill, which is much longer in proportion ;
and also by the middle, claw of each foot,
which are toothed like a saw, to enable it to
seize,

feize, and more fecurely hold its flippery prey.

So numerous is the tribe of herons, that Briffon has enumerated forty-feven different forts. Though exceffively voracious, they are always lean and hungry, weighing no more than about three pounds and a half each, notwithftanding they meafure three feet in length, and five in breadth. Although it is moft formidably armed with bill and claws, it is fo cowardly as to fly from a fparrow-hawk. Fifh and frogs are its chief food; but it cannot endure a long abftinence. Its voracity is fuch, that Willoughby fays, one of them will deftroy 15,000 carp in fix months. It lives among pools and marfhes, where it wades after its prey; and builds in the higheft trees, or on cliffs hanging over the fea.

The flefh of this bird, which is now thought difgufting, was formerly much efteemed. What an inftance is this of the capricious tafte of man!

Keyfler fays, that the heron very frequently lives to the age of fixty years.

The EGRET, or GREAT WHITE HERON.

THE length of this bird, from the bill to the claws, is four feet and an half, and to the

the end of the tail, three feet and a quarter; and the weight about two pounds and a half. It is entirely white, which diftinguifhes it from the common heron, which is rather larger, has a longer tail, and no creft.

The leffer white heron only differs in fize, and by having a creft.

The little white heron, according to Catefby, has a crooked red bill, with a yellow iris on the eyes, a white body and green feet.

To the above may alfo be added, the Yellow and green heron, found near Marfeilles; the bill of which is black above, yellow below, and about three inches long; the iris, as well as that part of the neck, next the chin, are white; but the reft of the neck, top of the head, the breaft and belly, are variegated with brown lines; the feathers on the back are black; the wings are yellowifh, fpotted with black; and the tail is ftuck with feathers greatly refembling hair. The thighs are of an afh-colour; and the feet are black, with yellow claws.

The LITTLE, or BRAZIL BITTERN.

THIS bird is fmaller than the common pigeon, although the neck is feven inches in length. The fkin, at the bafe of the bill, is yellowifh; the upper part of the head is of

a fteel

a steel colour, interspersed with pale brown feathers; the neck, breast, and belly, are whitish; and the back is a mixture of black and brown; the long feathers of the wings are of a greenish hue, with a white spot at each extremity; all the other parts of this bird, are beautifully variegated with black, brown, and ash-colour. The bill, which is long, straight, and sharp, is black at the point; the iris of the eyes is of a gold colour; and the tail is so short, that is does not extend beyond the wings.

To the above may be added, of the same species, the common bittern, the North-American bittern, and the small bittern.

The SPOON-BILL, or SHOVELLER.

WHO can behold this strange and singular bird, without adoring the wisdom of the great Creator of the universe! The bill of this bird alone, is a convincing proof of the great care of Providence to preserve his creatures. This bill is about eight inches long, and of equal breadth and flatness from one end to the other; but, contrary to that of all other birds, instead of being widest at the base, and narrowest at the point, is exactly the reverse, swelling into a broad rounded end, like the bowl of a spoon, from which it derives its name. It is, however, not hol-

low,

low, like a fpoon ; but whether clofed or open, it has a very fingular appearance.

This bird is as white as fnow, and, from its cleanlinefs, looks wonderfully pretty. It is common in Europe, and frequents the wa-ters.

The bill is moft peculiarly formed for the neceffities of this bird; as feeding principally on frogs, which, by their cunning and acti-vity, avoid the birds with pointed bills, the fpoonbill, by being notched and toothed all round, is better adapted, not only to take thefe animals, but alfo to prevent their efcape after they are caught.

The fpoonbill of America, is of a delight-ful rofe-colour, or beautiful crimfon.

The F L A M I N G O.

THIS bird is another inftance of the care of the Creator, in providing for every crea ture according to their refpective neceffities. Thus we fee the flamingo, which lives about the fhallow fhores of the fea, and the mouths of rivers, provided with a moft uncommon length of neck and legs; the latter of which are fo long, that when walking in the water, it appears as if fwimming ; and the head, which is almoft conftantly under water, in fearch of food, makes the bird feem no larger than a goofe, the body being then

only

only perceptible. But how great is the aftonifhment of the fpectator, when, on coming out of the water, it prefents itfelf, in height of legs and neck, like an oftrich! Its height is not only fuperior to that of any other bird, but its beauty is fcarcely to be equalled. The body is fnow-white ; the wings are of fo bright a fcarlet, as to dazzle the fight ; and the long feathers are of the deepeft black : the beak is blue, except the tip, which is black, and fo fingular in fhape, as to appear broken : the legs and thighs, which are not much thicker than a man's finger, are about two feet eight inches in length ; and the neck nearly three feet more ; the toes. are webbed, like thofe of the duck, which enables it to fwim for the prefervation of its life, which would be otherwife fometimes in danger, by the fudden rife of wind and water, while ftanding to a great depth in fearch of prey, by carrying it out to fea, where it might perifh for want of fubfiftence.

A difh of flamingo's tongues, Dampier fays, is a feaft for an emperor.

Flamingos always go in flocks, and are found in vaft numbers in Canada. Their nefts are formed of mud, refembling very much our chimney pots. When the female lays her eggs, fhe fits aftride the neft, with her legs hanging in the water.

" Thofe who admire," fays a learned writer, " the wonderful means, by which

" the God of nature has contrived, that
" thofe animals, which he has endued with
" a leffer principle than reafon, fhould pro-
" vide themfelves with food, and fecure
" their exiftence, during a life in which
" they are liable to innumerable accidents,
" would add a great deal to the meafure of
" their furprife, did they comprehend the
" variety of thofe means."

The AVOSETTA, or SCOOPER.

THE avofetta is diftinguifhed from all
other birds, by the bill, which turns up inftead of down, being about three inches and
a half in length, compreffed very thin, and
of a flexible fubftance, refembling whale-
bone. The tongue is fhort; the head, and
greateft part of the body, is black: the tail
confifts of twelve white feathers; the legs
are very long, of fine blue, and featherlefs
higher than the knee; the webs are dufky, and
very deeply indented.

Nature has fo peculiarly formed the bill
of this bird, to enable it to fcoup out of the
fand the worms and infects, on which it feeds.
It lays but two eggs, which are about the
fize of thofe of the pigeon, of a white colour, tinged with green, and fpotted with
black.

Thefe

Thefe birds are frequently feen, in the winter, on the Eaftern fhores of England; in Gloucefterfhire, tho mouth of the Severn; and fometimes on the lakes of Shropfhire. They have a lively chirping note, and very frequently wade in the waters.

The CURLEW.

THIS bird is, in length, from the top of the bill to the end of the claws, twenty-nine inches; and the breadth between the extreme points of the wings, when extended, is three feet four inches: the bill, which is nearly fix inches long, is narrow, a little crooked, and of a dark brown colour; the legs are long, bare; and of a dufky blue with a thick membrane meeting at the firft joint, and marked with irregular brown fpots.

This bird is of a greyifh colour, and the flefh very rank and fifhy, notwithftanding the Englifh proverb in its favour. They frequent fea coafts in large flocks, in the winter time, walking on the fands, in fearch of their prey, which confifts of crabs, and other marine infects. In the fummer they retire to the mountanous parts of the country, where they pair and breed.

The leffer curlew, called alfo the wimbrel, greatly refembles this bird; the chief difference

ence being in the fize, this weighing only twelve ounces, whereas the other weighs twenty-feven ounces.

The WOODCOCK.

THIS bird, which is fmaller than a partridge, is fourteen inches in length, twenty-fix inches broad, and about twelve ounces in weight. It has a ftraight bill, which is three inches long, the upper one falling a little over the under at the tip: it is of a dufky colour towards the end, and reddifh at the bafe; the forehead is afh-colour, and a black line extends from the bill to the eyes; the head, neck, back, and coverts of the wings, are irregularly barred with red, black, grey, and afh-colour; but, on the head, the black is moft predominant. The eggs are long, of a pale red, with fpots and clouds of a deeper colour. The flefh is reckoned a great delicacy.

In the fummer, the inhabit the Alps of Norway, Sweden, and other northern parts of Europe; but, when the froft commences, they retire to France, Germany, Italy, and Great-Britain.

Of

Of *WATER FOWL* in general.

THE principal diſtinction between land and water fowl, is, that the toes of the latter are webbed for ſwimming. Thoſe who obſerve the feet or toes of a duck, will eaſily conceive how admirably they are formed to move in that watery element, to which they are moſtly deſtined. What man performs by art, when he cloſes his fingers in ſwimming, the water fowl is ſupplied by Nature to perform. The toes are ſo contrived, that, when they ſtrike backward, the broadeſt hollow ſurface beats the water; but, as they draw them in again, their front ſurface contracts, ſo as not to impede their progreſſive motion.

The legs of the water fowl are generally very ſhort which cauſes them to walk with much difficulty ; they, therefore, ſeldom breed far from the ſides of waters, where they uſually reſort.

Thoſe of this glaſs, which have long legs are ranked among the crane kind ; ſuch as the flamingo, avoſetta, &c. which, although their feet are webbed for ſwimming, they ſeldom make uſe of for that purpoſe; a proof that their webbed feet are given them for the purpoſe of preventing their ſinking in the muddy ſhores, which they frequent in ſearch of their prey.　　　　　　　　　　　　We

We fhall felect a few of thofe moft worthy the notice of our readers, taking the pelican as the firft fubject for defcription.

The P E L I C A N.

TRAVELLERS, and thofe who are fond of the marvellous, have related ftrange accounts of this bird. The tale refpecting the care of its young, has been fo generally received, as to be frequently adduced as an example for man to imitate.

This bird is fo unwieldy, as to be only adapted for the water ; the beak, which is peculiarly uncommon, is about a foot long, and as thick as the flefhy part of a child's arm, very fharp at the point, and of a blue and yellow colour ; in other refpects, it differs very little from the fwan : the lower chap is made of two long flat ribs, with a rough membrane connected to both, in form of a bag, which, extending to the throat, holds a confiderable quantity of food, which fupplies it in times of fcarcity. Feeding her young from this bag, has fo much the appearance of feeding them with their own blood, that it caufed this fabulous opinion to be propagated, and made the pelican an emblem of paternal, as the ftork had before been chofen, more juftly, of filial affection.

O   The

The voice of this bird is harſh and diſſonant; ſome compare it to the braying of an aſs, while others ſay it reſembles the voice of a man grievouſly complaining. David compares his groaning to the pelican of the wilderneſs, and the owl of the deſert.

It lives ſixty or ſeventy years.

The FULMAR.

THIS bird is found in the iſland of St. Kilda, where it ſupplies the inhabitants with oil for their lamps, down for their beds, a balſam for their wounds, a delicacy for their tables, and a medicine for their diſeaſes. It likewiſe denotes a change of wind.

This bird is larger than the common gull; the bill is very ſtrong, yellow, and hooked at the end. Inſtead of a black toe, it has a kind of ſtraight ſpan. It feeds on the blubber of fat whales, and on ſorrel. It will leap and prey on a newly caught whale, even while alive; and is ſo voracious, as to eat until it is obliged, through repletion, to diſgorge its food.

Whales are frequently diſcovered by means of theſe birds, which collect together in vaſt numbers, and follow them, in hopes of prey, as ſharks follow ſhips that have diſeaſe on board, with the ſame expectation. The blubber on which they feed is what furniſhes

nishes them with the oil above-mentioned.
They seem, therefore, as if created for the
purpose of supplying the inhabitants of that
part of the globe with a commodity so essential to light them in those regions, which
could not otherwise be cheered from the win-
try gloom..

The GULL and PETREL.

OF these birds, the larger sort are most
shy, and live at the greatest distance, while
the smaller sort reside wherever they can take
their prey. They are principally distinguished by an angular knob on the lower chap of
the bill, which the petrels have not. The
sea swallow, which is also of this species, has
a straight, slender, sharp-pointed bill. In
their abodes and appetites, however, they all
agree, hovering over rivers, and preying on
the smaller fish, as well as following the
ploughman into the fallow fields, to pick up
insects. When they can find no other sub-
sistence, they will feed on carrion. They
are to be found in the greatest abundance on
our boldest rocky shores, where they find a
retreat for their young, in the cavities with
which those rocks abound. Like all birds of
the rapacious kind, the gull lays but few
eggs. It builds its nest, of long grass and
sea weeds, on the ledges of rocks. The flesh
of

of this fpecies of birds is black and ftringy,
and generally of a fifhy tafte; but that of
the gull is fomething better.   Of thefe, the
poor inhabitants make their fcanty and wretch-
ed meals.   Strangers to almoft every other
food, falted gull proves to them the greateft
dainty.   Thus we perceive that neceffity can
even create a comfort, by giving a relifh to
the coarfeft diet.

The TAME DUCK.

THIS is the moft eafily reared of all our
domeftic birds, the very inftinct of the young
leading them directly to their favourite element; nay, even when hatched by a hen,
which fometimes happens, they feek the water, contrary to every admonition of the fof-
ter-parent.

Of the tame duck, there are no lefs than
ten different varieties; but Briffon reckons
upwards of twenty forts of the wild duck.
The principal diftinction between the fpecies
is, that the tame duck has black, and the
wild duck, yellow feet.   The common fpecies of tame duck take their origin from the
mallard.

Ducks require very little charge in keeping,
living chiefly on loft corn, fnails, &c. for
which reafon they are very ufeful in gardens.
When they fit, they require no attendance,

except

except fprinkling a little barley, or refufe corn near them, which will prevent their ftraying.

Of the duck fpecies, there are alfo the eider, wild, velvet, tufted, pin-tail, grey-headed, white-bellied, Barbary, Madagafcar, and Bahama ducks.

Wild ducks are taken in decoys, and in fuch vaft quantities, that upwards of £.30,000 worth of wild ducks, wigeon, and teal, have been fent up to London in one feafon, from the decoys in the neighbourhood of Wainfleet only.

A DE-

A

DESCRIPTIVE ACCOUNT

O F

VARIOUS SONG BIRDS;

With PRACTICAL INSTRUCTIONS for chufing, breeding, feeding, and teaching them to fing.

ABERDIVINE.—This bird refembles, in
fize and colour, the grey canary. The cock
is diftinguifhed by a black fpot on his head,
and a little black under the throat; the hen
is greyer, with a fpotted breaft and belly.
They are both familiar, and eafily taken.

Food.—They love white feed; but are
moftly fed as linnets and goldfinches.

BLACKBIRD.—For the defcription, fee
p. 139.

Food.—When young, feed them every two
hours with frefh lean meat, minced very fmall,
and mixed with bread, a little moiftened.
When older, they may be fed with any raw,
or dreffed meat, if not ftale or four. They
fhould have water to wafh and prune their
feathers.

BULL.-

BULLFINCH.—This bird is in great estimation for its beautiful plumage, as well as singing, and also for its familiarity and tractability. It may be taught to pipe and talk, while perching on the finger, which renders it very engaging. To distinguish the cock from the hen, pull a few feathers from the breast, at about three weeks old, when those of the cock will be of a curious red, while those of the hen will be pale brown.

In order to teach this bird to pipe with propriety, a flagelet or bird organ should be made use of, while they are in the nest, and unfledged ; which, if properly attended to, they will retain a tune with the greatest exactness. Although the hen is not so beautiful in plumage as the cock, yet, with attention, she will very frequently pipe, and talk equally well with the male.

Food.—When young, give them rapeseed, soaked in clear water for eight or ten hours, then scald, strain, and bruise it, and mix it with an equal quantity of white bread, soaked in water, boiled with a little milk; it must be made fresh every day, to prevent its turning sour, and spoiling the birds. When they can feed themselves, give them rape and canary-feed, mixing most rape, as for linnets. If they droop, put a blade of saffron in their water.

CANARY BIRD.—This being the most estimable bird for its note, among those who delight

delight in finging birds, although of foreign origin, we could not avoid inferting a fhort account of it.

It derives its name from the Canary ifles, its original native country. Of the feveral colours, thofe which have white tails are the leaft valued. The mottled birds are thofe which are chiefly brought into this country by the Germans. The cocks are of a lively yellowifh colour, the hens of a dufky white.

To choofe a good canary, obferve that he ftands bold, ftraight, and upright, upon his perch; let his looks be fprightly, full of life and vigour; let him look freely at you, while looking at him, without fluttering or beating himfelf.

Food.—Give him, now and then, mawfeed, in which he principally delights, and fometimes a bit of loaf-fugar, between the wires of his cage; in warm weather, a little feedy chick-weed or groundfel. The fine leaf of a young radifh, heart of a cabbage, cofs, Silefia lettuce, or endive, will ferve to vary his food, which, being thus changed, will prevent his lofs of appetite, and ficknefs, caufed by keeping him on the fame diet.

CHAFFINCH.—The cock chaffinch, at about ten or twelve days old, has much white in his wings and pinions, with a reddifh breaft, and all his feathers higher, and
more

more brilliantly coloured, than thofe of the hen. An old cock has a blueifh head, reddifh brown back, mixed with green and afh-eolour, fine purple red breaft, and a white belly. The breaft of the hen is grey.

This bird is very docile and familiar, and may be taught, with attention, any tune; if put in eompany with other birds it will imitate their notes. The eock will couple with the Canary bird.

Food.—Rape and Canary feed.

GOLDFINCH.—This bird, whieh is greatly admired for fong and beauty, is the fineft feathered of all cage birds, and fo long. lived, that Willoughby mentions one. to have lived. twenty-three years. The eock is diftinguifhed by a curious fcarlet cirele round the fore. part of his head, or bafis of the bill.

Food.—When young, give them white bread, foaked in elean water, to a very thiek confiftence. To. this, add a little flour of Canary-feed. They fhould be fed at leaft every two hours, but very fparingly, and with frefh food every day. In about a month, you may wean them gradually from this foft food, by laying fome Canary-feed befide, until they can be brought to live on it entirely.

GREEN-FINCH, green-linnet, or green-bird, is of a hardy nature, and rather larger than the chaffineh. The head and back of the cock are green, edged with grey. The
middle

middle of the back inclining to chefnut. The fore part of the head, neck, breaft, quite down the belly and rump, are of a yellow green.

Food.—The fame as the chaffinch.

Common LINNET.—This bird is faid to excel all the fmall Englifh birds in finging. The note is curious; and he can imitate the fong of any other bird. The cock has a browner back than the hen, and more white in its wings. When the wings are full grown, fecond, third, or fourth feather, is white up to the quill.

Food.—They fhould be fed with feed gathered from the land where they are taken, mixed with a little bruifed hemp-feed.— When caged, give them a fmall quantity of Canary, and a few corns of hemp. If drooping, a little lettuce-feed, and a fmall piece of liquorice or faffron put into their water. Chick-weed is alfo a great reftorative to the linnet.

NIGHTINGALE.—The nightingale is reckoned the beft of fong-birds. In grown birds, the cock is diftinguifhed by its deeper and higher colours. In neftlings, when he has eaten, he gets upon the perch, and begins to tune to himfelf.

Food.—Give him, three times a week, two or three meal-worms, or fpiders, to purge him. When his fat declines, give him

him a little faffron in his water. Figs, chopped fmall among their meat, will recover their flefh when very thin.

RED-POLE.—This bird is very prettily feathered; the head and breaft of the cock being of a fine red, and much more brilliant than thofe of the hen. It is not much efteemed for its finging, although it has rather an agreeable note. Its neft never being found in England, denotes it to be a foreign bird.

Food.—The fame as the linnet.

RED-START.—The cock is a very beautiful bird. The tail, rump, and breaft, are of a fine red. The back, neck, and hind part of the head, are of a lead colour. The throat, and fore part of the head, are jet black, and it has a white mark on the pole. He is diftinguifhed moftly from his black head. He doubles his notes very finely, and will fing in the night as well as the day.

Food.—The fame as the nightingale.

ROBIN-RED-BREAST.——This bird, which is naturally folitary, will, when impelled by cold, become daring, familiar, and fociable. The red on the breaft of the cock is deeper, and extends farther upon the head than that of the hens. His legs are alfo darker, and he has generally a few hairs on each fide of his bill.

Food.

Food.—The fame as the wood-lark, or nightingale, but be careful not to overcharge their' ftomachs. Never let them want frefh water, and once a week, put in it a blade of faffron.

SKY-LARK. At about a month old, the cock may be known by his notes, which, though low, are diftinctly altered. In old birds, the cock is the lighteft coloured, has a browner back, a yellower throat and breaft, and a white belly.

Food—Give them egg, bread, and bruifed hemp-feed, with red fand at the bottom of the cage, and they will grow tame in two or three days. The neftlings fhould be fed, every two hours, with white bread and milk, mixed with one third part of rape-feed, foaked, boiled, and well bruifed. A fheep's heart, or other frefh meat, minced fmall, is good for them; and, now and then, they fhould have a hard egg chopped very fine, an equal quantity of hemp feed bruifed, and a little bread grated among it. Give them a turf of three-leaved grafs twice a week to perch upon.

SPARROW. The hedge-fparrow may be tamed fo as to fly about the houfe, without any apprehenfion of its ftraying. It will take the fong of the beft finging birds, if properly placed with them. The cock has a long, flender, dufky coloured bill. The up-

P   per

per fide of his body is black, mixed with a dirty red, and the breaft is black.

Food.—When taken, feed them, at twelve days old, with minced frefh meat and bread, or woodlark's meat. When brought up, give them hemp and Canary. If drooping, mix it with a little oatmeal.

STARLING. Having defcribed the ftarling in page 140 of this volume, we have only to obferve that their food is the fame as that of the blackbird, or woodlark.

THRUSH *or* THROSTLE. The thrufh has a great variety of notes, and fings nine months in the year. The feathers of the cock differ from thofe of the hen, in beauty, fleeknefs and brilliancy.

Food.—When full grown, feed them with frefh meat, raw or dreffed, with bread. This agrees beft with them, though they may be brought to feed entirely on bread or hempfeed. They fhould have a frefh pan of water twice a week. When cramped, put fern or clean ftraw on the bottom of the cage, and feed them, as they lie, with nightingale's meat.

TITLARK. This bird is handfomely fhaped, and excelled by very few. It has no remarkable fong, unlefs the cock is particularly excellent, when it will fing like a Canary bird. The neftling cock has more yellow

low, efpecially under the throat, legs, and
foles of the feet, than the hen.

TOM-TIT, otherwife Joe Bent, is a very
pleafing bird, and has a pretty fong.
Food.—They will thrive with bread and
cheefe, and, when grown up, with hemp-feed.
But they relifh the wood-lark's food the
beft.

TWITE. This bird; which is fuppofed
to be a native of Germany, vifits England
in winter. It is very brifk, and always finging. It is gentle, familiar, and is hung
among other birds, to provoke them to fing.
The cock is known by a red fpot on the
rump.
Food.—Rape and Canary : but they like
the latter beft..

WOODLARK.. The woodlark is efteemed the beft fong-bird in Great-Britain. It
fings nine months in the year. The cock is
known by its fize and fong.
Food.—Hard egg, chopped and minced,
with crumbs of bread, a little hemp and
maw-feed. One egg is enough for fix larks.
Give them fometimes minced meat, as other
birds, but no turf in their cage.

WREN. This is the fmalleft of fong-
birds, being about four inches long, from
the top of the bill to the end of the tail. It
has,

has, however, a very loud fong. The cock
has a dark brown back and head, with a
white breaft and bill; the tail and wings are
of a bright yellow, variegated with dark
lines.

Food.—The fame as the nightingale; but,
when fick, two or three flies, or fpiders.

NATURAL.

NATURAL HISTORY.

INSECTS.

THEIR GENERAL NATURE.

DEFINITION.—Infects are fmall animals, breathing through vent-holes, arranged along their fides, and provided with a fkin, of a bony nature. Their body is compofed of a head, trunk, limbs, and abdomen.

Form and ſtructure.—Not having occaſion to fly far, they are not made fo fharp before as birds: but their wings have fuſficient ſtrength and activity to conquer all the refiſtance they meet with, in their ſhort paſſage through the air. Having neither bones, fleſh, nor fkin, as in other animals, they are covered with a curious coat of mail, which both guards and ſtrengthens the body, while it renders the infect more adapted to the pur-

poſes

poſes of ſeeking its food, and performing every other function of its being.

Eyes and antenne.—The eyes of the fly tribes are two little creſcents, or immovable caps, around the head of the inſect ; and contain a great number of minute eyes, ｜croſſing each other in the form of lattice-work. Curious obſervers relate that they have counted ſeveral thouſands in each combination. Lewenhock calculated as many as 8000. The cauſe of their eyes being ſo numerous, is to ſupply the defect of viſion ariſing from their eyes being immovable. Thus inſects have eyes in every direction. How admirable muſt their ſight be, which enables them to diſcern objects, with their innumerable quantity of eyes, with as little confuſion as other animals do with two ! Their antennæ are ſmall horns projecting from their head, in ſuch a manner as to preſerve the ſight of ſo many fixed eyes from being injured.

Motion.—The admirable mechaniſm in theſe that creep, the curious oars of thoſe that ſwim, the incomparably formed feet of thoſe that walk, the ſtrength and elaſtic force of thoſe that leap, and the talons of thoſe that dig, afford the moſt ample matter for contemplating the endleſs wiſdom of the Creator. Each is particularly adapted to the kind of motion peculiar to the reſpective inſect; which is exemplified in the graſshopper, water-beetles, crickets, &c. To render their progreſs through the air as eaſy as

poſſible,

poffible, infects are provided with wings, formed of the lighteft membranes, and the fineft articulations. To poife the body, fome 'have four wings ; while fuch as have only two, have pointels, or poifes, under each wing.

Parts.—Infects are compofed of joints, mufeles, tendons, and nerves ; with eyes, brain, ftomach, entrails ; and with every other part of an animal body. How is the mind abforbed in wonder, when it confiders that the fmalleft animalcula, which the microfcope can only render vifible, is poffeffed of all the above related parts ! May we not, therefore, fay with Galen, when fuch exqui-fite workmanfhip appears in the minuteft in-fect, what muft be the wifdom employed by the Almighty in forming the more noble parts. of the creation ?

Sagacity.—Whether by inftinct, or actu-al fagacity, infects are fecured'againft winter, our admiration is equally raifed. When cold and wet oblige them to retire, fome en-tomb themfelves, as in their Aureila, or chryfallis ftate; others provide themfelves. in fummer with fufficient provifions for. their-winter fubfiftence; and fome of the infect tribe exift in a fleeping ftate, without chang-ing their nature, or being under the neceffi-ty of requiring that food which is denied them by the change of feafon. This caufed Solomon moft wifely to fay, " Go to the ant, thou fluggard, confider her ways and

be

be wife; which having no guide, overfeers
or ruler, provideth her meat in the fummer,
and gathereth her food in the harveft."

Care of their young.—Infeéts, with the
greateft care and affeétion, carry their young
in their mouths, which is particularly obferved in the ant tribe. But their care, in
general, deferves the greateft admiration.
They depofit their eggs in fuch places as fecure, produce, and fubfift their offspring.
According to the fpecies, their eggs are
laid in waters, on woods, or on vegetables,
where the young find a fubfiftence agreeable
to their nature. Particular woods, herbs
and plants, are chofen by the parent infeét to
fofter their future offspring. Thus nettles,
ragwort, cabbage-leaves, oak-leaves, currant
and goofeberry bufhes, &c. have their peculiar infeéts. Some, whofe eggs require
more warmth, depofit them in the hair of
animals, the feathers of birds, and even in
the fcales of fifhes. Others make their nefts
by perforating earth and wood, where they
depofit their eggs with fuch neatnefs as to
gratify the moft curious obferver. And to
prevent their eggs being injured, they inclofe
them in the leaves of vegetables, curioufly
glued together.

Food.—Every fpecies of infeét has a food
peculiar to itfelf. Caterpillars, for inftance
are not only limited to herbage, but, likewife, to a peculiar kind. Sooner than difo-
bey this ordinance of Nature, they will perifh
with

with hunger, unlefs they meet with a plant
fimilar to that to which they are attached.
To this general rule, we admit there are
fome few exceptions in caterpillars that will
fubfift on any vegetable. This feems to be
wifely regulated, in order to prevent the
moft ufeful parts of vegetation being deftroyed by caterpillars feeding, for inftance, on
apple-trees only.

Ufe.—Let no perfon confider the infect
part of the creation, as only worthy to be
crufhed to death by the foot, or to be made
the cruel fport of thoughtlefs childhood :
for, in the words of the, ingenious and im-
mortal Shakefpear, " The poor beetle,
" crufhed beneath the foot, feels a pang as
" great as when a monarch falls." Surely
their weaknefs ought to be their fureft protection againft fuch treatment. But, when it
is confidered that we derive the greateft embellifhments, and medicinal aids, from their
virtue, felf-intereft, if not gratitude, fhould
protect. their defencelefs lives from being deftroyed by man. · To them we are indebted
for our filk, honey, cochineal, and feveral medicines that are indifpenfibly neccf-
fary to preferve our lives from being the
prey of maladies that might otherwife prove
incurable. Added to this caterpillars are
indifpenfible food for birds, in their infancy,
which have then their cries heard and relieved by the Creator, producing this fubfiftence, fo admirably adapted to their tender
texture.

texture. But fometimes it muft be allowed, that the Almighty punifhes the ingratitude of man, by fending hofts of flies, locufts, and caterpillars, in array againft him. This fhould teach us not to defpife even a worm, which has been fo frequently rendered one of our moft powerful and dreadful enemies. Let us not think ourfelves rich, great or independent, while the Almighty can punifh our prefumption with fo inconfiderable an inftrument.

Tombs—The caterpillar, fatiated with verdure, retires voluntarily from life, and feeks the grave. Previous to their retreat, they change their fkins, ceafe to feed, while they build themfelves a tomb, or fepulchre. A few days conduct fome of them into a new ftate, of fuperior exiftence. Inftead of crawling the earth, they wing the air. The intermediate ftate between the worm and the fly, and which is fo ftriking a picture of diffolution, is called the cryfallis ftate. What appears the tomb of the worm, is the embryo of the butterfly; which, here acquiring a perfect form, burft the barriers of the grave, and fpeeds its flight into another world of enjoyment. What a contraft of being is there between its laft and former ftate! The caterpillar is terreftrial, and crawls heavily along the ground. The butterfly is agility itfelf, and feems almoft to difdain repofing on the earth, from whence it derived its being. The firft is fhaggy and of hideous afpect; the latter.

latter is arranged in the greateſt ſplendor and beauty of glowing colours. The former was obliged to a groſs food ; but this imbibes the eſſence of flowers, regales on dews and honey ; and perpetually varies its pleaſure, in the full enjoyment of nature, which it moſt delightfully embelliſhes.

A collection of theſe beautiful and variegated inſects is a ſplendid ſpectacle, where the richeſt and moſt diverſified colours delight and aſtoniſh the eye with their ſhade and diſpoſition. The ſight alone enraptures. But, what a ſublimity of reflection they afford to the contemplater of nature ! The period of the caterpillar's reptile exiſtence being accompliſhed, it entombs itſelf, for the purpoſe of riſing again a ſuperior being. The chryſallis is, at once, the tomb of the caterpillar, and the cradle of the butterfly. Under a tranſparent veil, this miracle of nature is effected ; from whence, like the ſons of man riſing from the tomb at the day of reſurrection, the butterfly breaks the barrier of its grave, and wafts itſelf into the air of heaven. Here it enjoys the effulgence of light and reſpires the breeze, embalmed with the ſweets of nature. Succeſsful in his rifling every nectarous flower, his reſt is the harbinger of enjoyment. His airy wings convey him from pleaſure to pleaſure, while they captivate man with their beauteous and variegated ſplendour. And in this revelling from eſſence to eſſence, he is not to be caught but

but by a fmall net of gauze, or filk, upon a
wire, placed at the end of a light wooden
handle.

What a fcene of wonders does not the butterfly difplay ! Its eyes of net work ; its
wings befprinkled with a farinacious duft,
of which every grain is a tile laid over a very fine net of gauze ; and the infinite variety
of form, colour, richnefs, and beauty, of
its embellifhments, render it fo wonderful,
that the ladies of China are faid to fpend
their whole lives in the ftudy of this incom-
rable infect. They inclofe, in a box filled
with fmall fticks, a number of caterpillers,
ready to fpin their bag ; and when they hear
the fluttering of the butterflies wings, they
releafe them into a glazed apartment, filled
with flowers.

In order to give our young readers as clear
an idea of infects, in their worm and cater-
piller ftate, as the limits of our plan will
allow, we have felected fix as the moft
beautiful and curious we could find, in
Dr. Lifter's Latin treatife on this part of a-
nimal nature.

The

THE ingenious Mr. Lifter fays, that, after he had fupplied this caterpillar with various kinds of herbs, which it was tired of eating, he has placed before it fome nettles; fuppofing it might be pleafed with a different kind of food. He faw, with great admiration, that the infect became fo joyous as to feem, by its motion, to congratulate itfelf on fuch a repaft being fet before it. But, fuch was the avidity with which the nettles were eaten, that not any remained of them in a very fhort time. Having thus nourifhed itfelf for a few days, it began in October to prepare for transformation. Being then put under a glafs, the infect affifted itfelf to the centre, and thus hung fufpended. Having attained the ftate of transformation, it fo ftrongly moved itfelf, and ftruck the glafs with fuch force, as even to caufe the vibration of the noife to laft while forty was counted. On the 12th of December, the fame author obferves, that a perfect infect was produced, which was exceedingly beautiful, and refembled in variety of colours the Peacock. It lived forty days; in which time he fays that he knew not any food on which it fubfifted.

Q   *The*

The GREEN MARBLED BUTTERFLY.

WHEN the coleworts and cauliflowers begin to heart, the perfect insect of this caterpillar is chiefly found depoſiting her eggs upon the leaves. The heat of the ſun ſoon vivifies the eggs, and brings forth the ſaid caterpillars, which immediately begin to conſume the vegetables above mentioned. They bear the heat of the ſun very eaſily : but they cannot endure long rains, and frequent ſhowers ; for in ſuch weather they waſte ſo faſt as, in a very ſhort time, to have no more remaining of their being, but the ſkin.— This worm begins to purge itſelf and prepare for its transformation, about the 3d of Auguſt ; and on the ſeventeenth of the ſame month the butterfly is produced. This perfect infect is very inactive, and ſlow in its motion. It however generally exiſts during the winter : and ſometimes it has been found alive when the ſpring has been far advanced.

The YELLOW UNDER-WINGED MOTH.

THIS kind of infect is of all the moſt difficult to be obtained. Liſter ſought in vain, a conſiderable time, to find in what place and
manner

manner it depofited the eggs. After many trials and enquiries, he placed one upon a leaf, which he had no fooner done, than it began to cover itfelf with a woolly fubflance, feemingly as a prefervation againft wet or cold. The leaf, being in a little time opened, he found a green feed : and he found that the infect fed on goofeberry-leaves, or curling vines ; and alfo the leaves of white, black, and red currants. It began about the end of June to prepare for its ftate of tranf- formation, in which it remained until the 13th of July, when a butterfly, fpotted with black and white, fprung forth, to enjoy its new ftate of perfect being. When touched, or fuffered to fall, it remained fo motionlefs as to appear entirely dead.

The NUT-TREE MOTH.

THIS worm, or caterpillar, delights in rofe-leaves ; but they are not fo ravenous as others : for they have long intervals between their meals. They feldom change their leaf until it is entirely confumed. Their colour is very elegant. The upper part of the bo- dy is of a beautiful yellow. But they are not fo beautiful after, as before feeding ; for their fkin is fo thin as to be tinged by the colour of whatever food they eat. Before it difpofes itfelf for transformation, the body

assumes

affumes a red colour.   This infect was found
to commence its aurelian ftate about the beginning of June; and on the 5th of December a perfect infect was brought forth.

The TIGER MOTH.

THESE caterpillars feed on the leaves of
red rofes, and red goofeberry-bufhes.   Some
have their feet in the middle of their body,
and others at the extremities. · When they
change place from one fituation to another,
they afcend by attaching themfelves to the
bough, with their feet, by which they raife
the body like a ferpent, and thus gain their
defired fituation. They hold themfelves fo faft
by their feet that they can fcarcely be taken
from the part to which they adhere.   They
prepare for transformation by cleaning their
bodies ; which being done, they commence
their chryfalid flate about the firft of April,
and on the 24th of July the perfect infect is
produced.

The

The PHOBERAN.

THIS caterpillar is found near a village called Groed, in Flanders. It is generally feen fitting on a branch of willow. It feeds on the leaves of the fame tree. It eats very leifurely. The hinder part of the body refembles the beard, face, and head of a goat. When you take it, it ftrikes' as if in the greateft anger. It has two hooks on the back, with which it guards and preferves itfelf from the attacks of other creatures. It is therefore called by Lifter, the phoberan. When it eats, the head appears tied to the body, with a flight thread, or filament, not unlike the joining of the head and body of a fpider.

On the firft of September, it refigns itfelf to its approaching transformation. Twenty-two days after, appears a beautiful butterfly, diftinguifhed for its beauty and variety of colours. Before the perfect infect, it depofits its eggs, which are coloured with different green hues.

Q 2   SERI-

WITHOUT entering into the description of a naturalift of this worm, we fhall confine ourfelves to that which we think will be more ufeful, pleafing, and interefting. It being more an object of univerfal fervice, than of fingular beauty, induces us to prefer giving an account of its utility, than any elaborate account of its figure or colour.

Where thefe worms are bred, they no fooner leave the eggs than they are fed with mulberry-leaves, with which they are fupplied every morning, when the old leaves are carefully removed. This infect, when firft produced, is extremely fmall, and entirely black. In a few days it affumes a new habit; which is white, tinged with the colour of its food. And before it goes into its chryfalid ftate, it affumes two other dreffes. At this time, it appears difgufted with the world, and voluntarily retires to its folitary grave, which is moft admirably formed with its thread. How wonderful muft be the ftructure of its body, to furnifh fuch a thread; and how aftonifhing the inftinct which teaches it to make, of this felf-produced material, its own tomb! And how muft it diminifh the pride of man, to confider that he is indebted, for his moft gaudy array, to a fubftance, of which a

worm

worm forms its fepulchre! Reflect on this, ye potentates of the earth; and acknowledge, with humble gratitude, your debt to the filkworm; and diveft yourfelves of the vain arrogance you affume when arrayed in the robes of majefty!

When the cryfalid ftate begins, the infect proceeds to fpin its filk, in which it is buried. Like the pierced iron plates of a wire-drawer, this worm produces the thread through a pair of holes in an inftrument placed under its mouth. Two drops of gum ferve it as diftaffs, fupplying the fubftance of which fhe fpins the thread; for the gum is no fooner in the air, than it lofes its fluidity, and changes to the filk, in the due fize of which the worm is never deceived. She always proportions her thread to the weight of her body. The cone of filk being formed, and opened, is found to confift of the worm, changed to a nymph, and buried in its centre, or down or flue, which is the bad part of the filk, and the perfect part, all ranged with great compactnefs and propriety. It may be a matter of wonder how fo fmall a moth as this little worm muft neceffarily produce, fhould be able to burft the million fold barriers of her place of regeneration.

The fame omnifcient being who taught it how to erect this place of reft, taught it, at the fame time, to find an eafy accefs to her aerial exiftence. The new animal, with its horns, head, and feet, directs its efforts to

that

that end of the cone it has left purpofely light enough to admit its paffage to another world of enjoyment.

By calculation, one of thefe worms will produce between nine hundred and a thoufand feet of filk at one fpinning : and fo thin and light is its texture, that the whole weighs no more than 2 1-2 grains. And as they were particularly formed to furnifh mankind with a fubftance for drefs, that might render us more agreeable to each other, and thus inhance the few pleafures of our exiftence, nature has caufed one fly to lay as many as 500 eggs. How grateful, then, we ought to be to the Creator, who thus forms, yearly, fuch an infinity of thefe manufacturers of the moft agreeable and beautiful fubftance the world affords, for our array and embellifhment ! By this worm, grandeur is more ennobled, and even royalty itfelf is rendered more majeftic.

THE

THE FIRST ORDER.

Infects with cruftaceous elytra covering the wings.

Genus I.

SCARABÆUS——*The* BEETLE.

ALL infects having wings covered with the elytra, or cafes of the wings, were ufually called in Latin, Scarabæus; until Linnæus difcriminated them, and confined the term to particular beetles, diftinguifhed by the horns on their head, and thorax or breaft.

SCARABÆUS AURATUS;

The GOLDEN BEETLE.

THE larva, or grub, of this infect, injures the roots of trees and plants. The beetle is found upon flowers, and particularly upon the rofe and piony. The whole is a burnifhed green, and tinged with red, fo as to refemble the fineft polifhed copper. The elytra are adorned with a few tranfverfal
spots,

spots, which add to the other embellishments of its brilliant colouring. Such is its amazing splendor, that it rivals the emerald, and is, therefore, admired as the moſt beautiful infect produced in England.

We avoid deſcribing the cockchafer, which, being fo well known, only requires us juſt to mention, that all its varieties depend on its mode of life; and its colours, on its ſex, age, health, ſickneſs, &c.

GENUS II.

LUCANUS——*The* STAG BEETLE.

THE ſtag beetle is the largeſt, and moſt fingular in its ſhape, of any in this country. It is known by two maxillæ, projecting from its head, and reſembling the horns of a ſtag. Theſe maxillæ are furniſhed with teeth, from their root to their point. The elytra have neither ſtreaks or ſpots. The whole infect is of a deep brown. It is ſometimes found in oaks, near London, where it is much ſmaller than thoſe of the ſame ſpecies found in woody countries. As their horns pinch ſeverely, they are carefully to be avoided. The greateſt beauty they poſſeſs is their maxillæ, or jaws, ſometimes appearing like coral.

The

The lucani feed on the oozings from oaks. Where the females depofit their eggs. The larvæ, or grubs, lodge under the back, or in the hollow of old trees ; which they bite, and reduce to fine powder. Here they tranfform themfelves into chryfalids.

The ufe of their porrected maxillæ, or jaws, is to loofen the bark to which they affix themfelves, while they fuck the juices oozing from the tree.

GENUS III.

DERMESTIDES.

Characteriſtics.

THE antennæ, or horns, end in a head of an oval form ; the thorax, or breaſt, is of a convex form ; and the head is fo bent as to lie almoſt concealed under the thorax.

DERMESTIS VIOLACEUS.

The VIOLET BEETLE.

THIS infect is exceedingly beautiful, and is much fmaller than, though nearly refembling, the ſtag beetle. The elytra are of a deep violet ; the thorax, or breaſt, is covered

ed with green hairs, and the legs are black.
The whole creature, glittering with its brilliancy, charms its observer. The larva and
the perfect insect being found in dead bodies,
evince that the Creator has power to produce
the most beautiful effects from the most disa-
greeable of mediums. How different is this
from human ability! With the choiceft of
nature's productions combined to almost infinity, man is not able to imitate the fplendor of this insect, which is produced by the
Almighty, from a dead and putrid body.

GENUS VII.

BYRRHUS SCHROPHULARIÆ.

The NETTLE BEETLE.

THIS insect is found mostly in flowers.—
Its oval body is black, except where the un-
derpart of the abdomen appears white, from
the multitude of minute scales with which
this part is covered. The elytra not only inclose the wings, but the sides and
under part of the body. These elytra are
black, with white and red scales, refembling
embroidery. This species is found in gardens. If rubbed, the small scales fall, and
cause the insect to appear entirely black.

GENUS

GENUS X.

COCCINELLA.

THIS genus comprehends thofe fmall beetles which have red and yellow grounds, fpotted with black; and are known even by children, who call them lady-birds.

Of all the different larvæ of the coccinella, the moft curious is that which, from its tufts of hair, and fingularity of figure, Mr: Reaumur calls the white hedge hog. It feeds on the leaves of trees; and having exifted a fortnight in its vermicular ftate, it turns to a chryfallis, without divefting itfelf of its fur; and, three weeks after, it takes flight from its tomb, as a perfect coccinella. When firft produced, the colours of the elytra are nearly white; but, in a little time, they change to that lively brilliancy for which they are fo juftly admired. Their eggs are oblong, and of an amber colour. This beautiful little infect is frequently found on thiftles.

GENUS XI.

CHRYSOMELA.

Character.

THE chryfomela have their antennæ, or feelers, fhaped like bead-necklaces. This genus contains a great variety of beautiful infects differing in fize, colour, and abode. They are found amoft every where, in woods, gardens, &c. When caught, they emit a difagreeable fmelling liquor.

———————

CHRYSOMELA GRAMINIS.

The GRASS CHRYSOMELA.

THIS beautiful infect, like moft of the genus, has an oval and very convex form. The colour is a fine gloffy green, fomewhat tinged with blue; which affords a moft charming reflect. The eyes are yellow, and the thorax and elytra are fpotted. It is found in the meadows, in May and June, upon water-betony, dead-nettle, mint, and other labiated plants. By fome it is called the blue-green chryfomela.

The

The glittering colours with which feveral fpecies of this genus, are embellifhed, difplaying the fplendor of, gold and copper, have conferred on them the pompous name of chryfomela. The larvæ prey upon the fubftances of leaves, without touching the fibres. The leaping chryfomela infeft the tender leaves of plants; which fhould be carefully guarded from their depredations.

GENUS XII.

THE antennæ grow gradually larger from each extremity to the middle, and are fituated between the eyes. The breaft and wing-cafes, are covered. Protuberant fpines.

HISPA ATRA.——*The* BLACK HISPA.

THIS pretty, fingular infeft; is of a deep polifhed black. The upper part of his body is clothed entirely with long and ftrong briftles, like the fhell of a chefnut, or rather in the manner of a hedge-hog. The cafe of the horns has even a thorn at its end, to guard the infeft from injury. The breaft has a row fet tranfverfely, which are forked. And the elytra, or wing cafes, are covered with a great number that are fingle. The

points

points of all are firm and piercing. This infect was found by Barbut, in the month of
July, at the root of some long grafs, in a field
near Paddington. This flying hedge-hog, if
we may be allowed the term, is difficult to
be taken. It bears its antennæ erect before
it, as guardians of its progrefs through the
ærial element.

―――――――――

GENUS XVI.

CERAMBYX MOSCHATUS.

The NUTMEG CERAMBYX.

THE body of this infect is entirely green,
tinged with blue and gold colour, which renders it moft delightfully refplendent. It is
fometimes found compofed entirely of blue
and gold. The elytra are long, foft and
flexible, and finely fhagreened. This beautiful creature is found upon the willow,
which it perfumes with an odour like that of
a rofe, fo as to fcent a whole meadow.—
Thus, we perceive, that nature beftows on
this infect the moft grateful odour, to fupply
the want of thofe delightful fcents of which
meadows are deprived by the field flowers
being fhorn by the fcythe of the mower; for
it is obferved, this charming cerambyx is
produced in its perfect ftate about the general
time

time of making hay. What care does Providence take to accommodate man with a never-ceafing variety of delights, adapted to charm every fenfe!

Genus XVII.

L E P T U R A.

Charaƈter.

THEIR antennæ are fetaceous or briftly; the elytra diminifh in breadth towards the extremity; and the thorax is round and flender.

L E P T U R A A R C U A T A.

The RAIN-BOW LEPTURA,

VARIES in refpeƈt to fize, and is of a deep black ground, refembling velvet. The antennæ are of a bright yellow, and nearly as long as the body. The elytra are adorned with high flame-coloured crofs bars, which are formed by a down of a moft refulgent golden yellow. Viewed through the microfcope, it appears like velvet inlaid with topazes; and, when affifted with the folar rays, nothing can excel its infinity of fplendor. This moft wonderful infect for beauty is the poor tenant of a decayed tree, on which it may be frequently found, efpecially on an alder.

The larvæ are found with thofe of the preceding genus, which they greatly refemble in appearance and mode of exiftence.

CASSIDA,——*The* SHIELD BEETLE.

THIS genus, which Barbut ranks under the ninth clafs, is thus named, from concealing its head under the margins of the thorax, as if it were defended with a helmet. Many of this fpecies are found in foreign countries. Their larvæ form for themfelves a kind of umbrella, which fhelters them from the fun and rain. Thefe infects inhabit thiftles and knotty plants. One fpecies of them produce a chryfallis, refembling an armorial efcutcheon. This brings forth that fingular caffida, which is fo diftinguifhed for its variegated beauties. Many are found upon the wild elecampane, growing on the fide of ponds.

GENUS

GENUS XIX.

L A M P Y R I S.

Character.

THESE infects are chiefly diftinguifhed by their emitting a light in the dark; and are, therefore, called fire-flies. The females are apterous or without wings.

L A M P Y R I S N O C T I L U C A.

The GLOW-WORM..

CONTRARY to the general order of na- ture, the male of this infect is lefs than the female. But the greateft difference betweer the fexes is, the male being covered witl brown elytra, fhagreened and marked witl two lines longitudinally: The two laft ring of the abdomen are not fo bright as thofe (the female, but they have four luminor. points.

The glow-worm, which is frequently fee in woods and meadows at night in June, i the female. The fhining light it emits di rects the male to his tender partner, whicl not being able to fly, is thus moft wonde:
                                                    fully

fully provided by Providence with a felf-pof-
feffing ray, in the fun's abfence, to fhew its
mate the fpot where it is anxioufly waiting
its company. Thus are the banks and hedges adorned with their little illuminations,
while the nightly traveller is charmed with
their beauteous fplendor.

Their luminous power depends on a liquor
placed at the lower extremity of the infect,
which by fuction renders it more fhining, or
by dilating or contracting itfelf withdraws
or emits it at pleafure. That the light is
caufed by a fpecies of phofphorus, is evident,
from the animal, when crufhed, leaving upon the hand a luminous matter, which continues its luftre until it is dried.

The perfect infect flies in autumn even-
ings, and frequents plantations of juniper-
trees.

━━━━━━
━━━━━━

The FIRE-FLY of the Eaft-Indies:

THIS fly is about an inch long, and an inch
broad. Their head is brown, and has two
fmall horns, or feelers. They have four
wings. On their backs, they have a black
bag, containing a luminous fubftance, which
is concealed by their wings, unlefs expanded
during their flight. In rainy feafons, they
fwarm among trees, and feed upon their
bloffoms. Of thefe flies, there are feveral
species

fpecies in the Eaft-Indies. Being deftined, feemingly to roam by night, in order to avoid the exceffive heat of the fun by day in thofe fultry climates, how providentially Nature has accommodated them with a fubftance that renders their ærial courfe perceptible to each other! But when they alight, and fwarm upon trees, their luminous fubftance, being no longer ufeful, is concealed and preferved by their clofed wings.

LAMPYRIS NOCTILUCA of *Martinico.*

The FIRE-FLY *of Martinico.*

THIS fly, according to the Pere de Tertre, is lefs than the common fly. They emit a fparkling golden light, which is extremely agreeable. But the infect withdraws, and lets it fhine at intervals, alternately, throughout the night. This effulgence is contained in a whitifh fubftance, of which the infect is fo full, as to make it appear through the crevices of its fkin at its pleafure.

Thefe different fire-flies feem deftined by Nature not only to chear the bofom of darkfome night, but to guide the wandering favage through the pathlefs wood, or defart wild. Indeed by their light, he may lay more fecret fnares for his fhaggy prey on the

mountian

mountain, or his finny prey in the deep, than he could by the presence of the fun.—Thus, being deprived of that artificial light which he can only possess from civilization, Nature has fortunately created these admirable insects for his convenience.

Genus XX.

C A N T H A R I S.

Character.

THEIR horns or feelers are bristly; their breast is margined; and their elytra, or wing-cases, are flexible. They are commonly called Spanish flies; but this is erroneous, as they are a distinct genus from the cantharides.

CANTHARIS LIVIDA.

The LEAD-COLOURED CANTHARIS.

THIS insect varies in the colour of the elytra; but this difference only arises from the difference of sex. Their horns are all black, except the articulation near the base, which are yellow. They have black eyes;

and

and the head, in both fexes, is a yellowifh red. The wing cafes are filky, flexible, and appear as if ftrewed with filver-duft, when viewed by a magnifying-glafs. The abdomen, or belly of this fly, is black; except the laft rings, which are yellow. It is found upon a flower.

CANTHARIS PECTINICOMIS.

The COMB-HORNED CANTHARIS.

THE antennæ, or feelers of this fly, are black, combed, and as long as the body. The breaft and elytra are of a beautiful fcarlet. It has black legs, and yellow eyes. It is a pretty infect, and is found among flowers.

This genus contains a number of beautiful infects, the colours of which vary according to the difference of fex, feafon, &c. which renders it unneceffary to defcribe them.—They frequent flowers;, and their larvæ are fimilar to thofe of the cerambyces, and are to be found in the trunks of decayed willows, and other old trees. Although thefe infects are frequently confounded with the cantharides, yet they differ effentially: for the canthares have five articulations in the tarfi, or intermediate part between the leg and foot;. but the cantharides have five articulati articulations or joints, only, on the two firſt pair of legs, and four only to the tarſi of the laſt pair.

Genus XXI.

The SKIPPER.

Character.

THEIR horns are briſtly; and they have an elaſtic ſpring, or ſpine, which projects from the hinder extremity of the breaſt.

ELATER SANGUINEUS.

The BLOOD-COLOURED SKIPPER.

THE breaſt of this inſect-ends, under-neath, in a long point, or ſpine, which enters, as if with a ſpring, into a cavity in the upper part of the under ſide of the thorax. By this admirable conſtruction, the ſkipper is enabled, when upon its back, to leap in the air, and, thus, alight on its feet. It varies in ſize; and, when young, the elytra are of a beautiful red: but in a few days they loſe this ſplendid hue, which is then changed to poliſhed black; and, when view-

ed

ed through a microfcope, to nearly a chef-
nut-colour. The breaft is a glittering, and
appears with dark down, interfperfed with
fome black hairs. The female is black, and
marked with fpots of a deeper die, occafioned
by a velvet down, lying in tufts, which are
only to be diftinguifhed by the glafs.

The larvæ are found in the trunks of
decayed trees, where they are transformed
into perfect infects, which flutter upon flowers, wander over fields, and conceal themfelves
in thickets, or under the bark of trees.

Genus XXII.

CICINDELA.

Charaƈer.

THE horns are briflly ; the jaws porrected, and armed with teeth ; the eyes are
prominent ; and the breaft is rather round,
and margined.

CICIN-

CICINDELA CAMPESTRIS.

The FIELD-SPARKLER.

THE field-fparkler is one of our moft
beautiful infects.   The upper part of its
body is rough, and of a fine green, tinged
with blue.   The under fide, legs and horns,
are of a fhot colour, gold, and a red,
inclining to the copper hue.   The eyes,
being prominent, give the head a broad
appearance.   The breaft is pointed, and
narrower than the head ; which characterizes
the cicindelæ.   Like the head, the breaft is
rough ; and of a green colour, tinged with
gold. The elytra are delicately and irregularly dotted, with fix white fpots on each.
This infect runs with great fwiftnefs, and
flies with facility. At the beginning of fpring,
it is found in dry, fandy places, where its
larvæ alfo inhabit.   Thefe are a long, foft,
whitifh worm, with fix legs, and a fcaly
head.   They make a perpendicular hole in
the ground, at the entrance of which they
keep their head, to catch other infects which
fall in it.   A fpot of ground is fometimes
entirely perforated in this manner.

The perfect infects of this genus are moftly
fo very beautiful, as to merit the attention
of the curious in microfcopic obfervations,
as well   as   in   natural refearches ; for
fome

ſome are minute, though not inferior in ſplendor to the larger; which renders them proper objects for the delightful amuſement of the magnifying-glaſs. And here it may be proper to obſerve, that living objects are always to be preferred to thoſe which are dead, by the enquirer into the produce of nature. The perfect inſects of this genus are, like their larvæ, perfect tigers in their diſpoſition for prey, which they attack, and deſtroy, with every effort in their power.

BUPRESTES GUTTALA.

The SPOTTED BUPRESTES.

THE whole body of this inſect is of a green and gold colour, with a blue tinge underneath. But it is chiefly diſtinguiſhed by four white concave ſpots upon the elytra. The entire upper part of this inſect appears moſt beautifully dotted, when ſeen through a microſcope.

The larvæ is ſuppoſed not to have been yet diſcovered: but from the ſimilarity of the perfect inſect with the clater, and both being found among timber and decayed trees, the larvæ and metamorphoſis may be imagined to correſpond.

CACABUS

CACABUS GRANULATUS.

The GRAINED BULL-HEAD.

THIS ſpecies is not only one of the largeſt, but the moſt beautiful and brilliant this country produces. The head, breaſt, and wing-caſes are of a coppery green. The elytra have three longitudinal rows of oblong raiſed ſpots. All the under part of the infect is black. But having no wings beneath the elytra, nature has providentially ſupplied it with ſuch legs as enable it to run with amazing ſwiftneſs. This infect is frequently found in damp places, under ſtones and heaps of decayed plants in gardens. The colour ſometimes varies ; for it is frequently found coloured with a beautiful purple.

The larvæ live under ground, or in decayed wood, where they remain until metamorphoſed to their perfect ſtate, when they proceed to devour the larvæ of other infects, and all weaker animals they can conquer.

They are frequently known by the name of the ground-beetle. Some are found ſo early as the beginning of March, in paths, &c. where the ſun warms the earth with his vivifying beams. Many of the large ſpecies have been found between the decayed bark and wood of willow-trees.

GENUS

GENUS XXVII.

MELOE.

Character.

THE horns refemble necklaces; the breaft is rather round ; and the elytra are foft and pliant.

MELOE VESICATORIUS, *or* CANTHARIDES.

The SPANISH FLY.

THERE are feveral fpecies of this infect, differing in fize, figure, and colour. But all are apparelled, by nature, with great luftre. Green, azure, and gold colours blend their hues to embellifh them. They are moftly natives of the fouthern parts of Europe. The fpecies ufed medicinally ·is nine or ten lines in length, of a fhining green colour mixed with azure, and very prolific. Thefe infects are fometimes obferved to fly in fwarms. A difagreeable fmell, like that of mice, indicates their approach. By this· fcent they are found by the gatherers, who collect them for the apothecaries. When

S 2 dried,

dried, fifty of them fcarcely weigh a drachm.
Shrubs, and particularly the leaves of afh-
tree, are their food. So corrofive are the
odorous particles emited by this infect, that
great caution is required in taking them.—
For many have been known to have fuffered
greatly, by only having gathered a quantity
of them with their bare hands in the heat of
the fun : fome. have been oppreffed with
fleep, by fitting under trees on which fwarms
of cantharides have fettled. Contrary to
the general cuftom of nature, the female
courts the male. The larvæ are produced
from the ground, where the eggs are always
depofited. Thefe infects, reduced to powder, are exceedingly efficacious as blifters,
in abforbing or drawing off humours which
threaten the effential parts of life. But the
cantharides is, notwithftanding, a moft formidable poifon, if taken internally without
the greateft caution. Some who have been
afflicted by their incautious ufe of them,
have found the beft antidotes to be milk,
olives, camphire, and oil of fweet almonds.

The larvæ of the meloes inhabiting this
country, greatly refemble the perfect infects;
for they are of the fame colour, are as large,
and are as flow in their motion. They are
generally found buried deep in the earth,
where they metamorphofe themfelves into
perfect cantharides.

We have introduced the meloe veficatori-
us, which is generally known by cantharides

rides or Spanifh fly to fhew in what it is different from a preceding-genus, called the cantharis, for which it is frequently miftaken.

CURCULIO, or WEEVEL.

THIS infect feeds upon corn, the infide of which it eats, and leaves the bran. In this tribe, nature difpenfes the riches of her moft refulgent colours, fo as to dazzle the eye with fplendor. But it is the microfcope that muft admit us to this fcene of fuperlative beauty.

The curculio regalis found in Peru is a wonderful inftance of the beauty nature can beftow on even what is generally deemed the moft inconfiderable of her products.

The larvæ, refembling oblong, foft worms, are greatly dreaded for the injury they do in granaries. Corn-lofts are frequently laid wafte by their ravages. The infect, having remained within the grain until it has devoured the meal, lies concealed under the empty hufk, until it paffes its aurelian ftate, and takes its flight as a curculio. While one fpecies feed on corn, others deftroy, in the fame manner, beans, peas, and lentils. To difcover the grain infefted by the larvæ, it is thrown into water, when that part which fwims is certainly perforated by the curcu-liones.

liones. The heads of artichokes and thiſtles are often deſtroyed by theſe deſtructive inſects. This animal being ſo delightful in appearance, and ſo deſtructive in its nature, is a leſſon which teaches that beauty may effect our ruin while it captivates our ſenſes.

<hr/>

GENUS XXX.

FORFICULA.

Character.

THE horns are briſtly; the wing-caſes are half the length of the wings, which, being folded, are, notwithſtanding, covered by the elytra; and the tail is forked.

<hr/>

FORFICULA AURICULARIO.

The EARWIG.

THIS ſpecies is entirely of a deer colour. The horns are prettily intermingled and variegated. The wings are of the ſame colour as their elytra, or caſes. This inſect is found in wet ſand, near pools and rivulets; and particularly on grape-vines. It is generally known, and dreaded by many for its

tendency

tendency to creep into the human ear. That
it has this habit, the editor of this volume
can affirm from experience : but, that perfons need be alarmed left it fhould, thus,
reach the brain, and caufe death, he denies ;
for the leaft acquaintance with the anatomy
of the head, will evince the impoffibility of
the infect reaching the inner part of the cra-
nium by the avenue of the ear, from there
being no communicate paffage from one to
the other. The forceps with which nature
has provided its tail, for defence, is capable
of biting, fo as to caufe, for the moment,
rather a painful fenfation. Although furnifhed with this defence, the earwig has been
obferved not to ufe it, even when he has
been furrounded with a fwarm of ants. But
it will frequently pinch the finger of perfons attempting to take them with their
hands.

The larvæ differs very inconfiderably
from the perfect infect.

THE

THE SECOND ORDER.

GENUS II.

MANTIS.

Character of the genus.

THE head is unfteady, and has a nodding motion. The mouth is armed with porrected jaws; and the antennæ, or feelers, are briftly. They have four wings, which are membranous, and wrap round the whole body. The firft pair of feet have teeth like a faw: and the breaft is narrow, and éxtends to a confiderable length.

MANTIS GANGYLODES.

The WALKING LEAF.

THIS infect is remarkably fhaped. The head is joined to the body by a neck longer than the body itfelf. It has two polifhed eyes, and two fhort feelers. The breaft is long, narrow, and margined. The elytra, which cover two thirds of the body of the infect, are veined, and reticulated, or netted. The wings are veined, and tranfpar-

ent

ent. The hinder legs are very long, the
next fhorter ; and the foremoft pair of thighs
are terminated with fpines. The reft have
membranous lobes, which ferve as wings
to them in their flight. The infeĉt might,
therefore, be juftly called the Mercury of
this part of the creation. The top of the
head is membranous, fhaped like an owl,
and divided at its extremity. This animal is
one of the innumerable inftances which nature affords, to indicate the infinite wifdom of the Creator. Whenever any part
of his workmanfhip is found to deviate from
the general fyftem, it is ftill formed to
anfwer the defign of its exiftence. This infeĉt, having fuch long legs, could never
have fuftained itfelf in the air, had not providence beftowed on it a fpecies of wings, to
balance its weight. Thefe are the inftances
with which nature teems ; and which would
make the atheift tremble, had he but fenfe
to contemplate the admirable defign, fyftem,
and application, with which they are cha-
raĉterized, as

———— parts of one ftupendous whole ;
Whofe body NATURE is, and GOD the foul.

This genus is generally of a very beautiful green ; but the colour foon fades, and
becomes that of dead leaves ; which has
caufed the inhabitants of China, where they
are found, to call them by the name of walking leaves.

The

The larvæ very much refemble the perfect infect: but it is feldom feen in this country.

Genus III.

Character.

THE head is bent inwards, armed with jaws, and furnifhed with palpe, or fpiral tongues. The wings are fo deflected as to wrap round the fides of the body. All the feet are armed with two crotchets, or nails; and the hinder are formed for leaping.

TETTIGONIA.—*The* GRASSHOPPER.

THIS infect walks heavily, flies tolerably, and leaps with wonderful agility. It has an inftrument in its tail, with which it digs holes on the ground, for the reception of its eggs. The grafshopper lays a great number at one fitting, of which they form a groupe, by uniting them with a thin membrane.— The little larvæ refemble entire the perfect infect, except in the fize, and having neither wings nor elytra. Thefe, as well as the perfect infect, are frequently found in meadows. They both feed on herbs very voracioufly.

cioufly. The grafhopper, having many ftomachs, has caufed feveral authors to affert
that they chew the cud, like fome other larger animals.

GRYLLUS,—*The* CRICKET.

THIS family of infects is called in England, crickets, from the found or noife they
make. Towards fun-fet they leave their
fubterraneous habitations, when they make
the fields refound with their chirpings. The
domeftic grillæ abide in ovens, and hearths
on which wood is burnt: here they frequently are troublefome, by their perpetual noife, and crawling about perfons fitting
near the fire. But a popular prejudice,
in many parts of England, prevents their
being · driven away, or deftroyed : for
poor peafants, and common people, imagine they bring good fortune to whatever
houfe they attach themfelves.—So true it is,
that the moft abfurd chimeras enter the
minds of the ignorant, who are always prone
to fuperftitious errors.

This infect is chiefly diftinguifhed by having at its hinder extremity two briftles.

The domeftic and the field cricket are the
fame fpecies; all the difference is, that the

former more inclines to yellow, and the latter to a brown hue.

Genus IV.

F U L G O R A.

Character.

THE front of the head is empty, and extended. The horns, which have two articulations, are fcaled below the eyes.

FULGORA CANDELARIA.

The LANTERN FLY.

THE head and breaft of this infect are generally the colour of a muddy brown; the elytra are of a lively green, fpotted with a pale yellow; the wings are of a beautiful yellow, and have their extremities bordered with a gloffy black. When the infect flies, the waving of the elytra caufes the tranfparent fpots to appear in the night like radient flafhes, forming various figures, according to the fancy of the wondering beholder. This fly is a native of China.

ANO·

ANOTHER LANTERN FLY.

THIS lantern fly is a nocturnal infect, that has a hood, or bladder, on its head, which appears like a lantern, in the night: but by day it is clear and tranfparent, and very curioufly adorned with red and green ftripes. Such a fhining light iffues from this part of the infect, that it is poffible to read by it. The wings, and whole body are elegantly adorned with a mixture of red, green, yellow, and other fplendid colours. The creature contracts or dilates the hood, or bladder, as it pleafes. When taken, they withdraw their light; but when at liberty, they fuffer it to fhine again, with all its wonderful refplendency.

Thefe flies are as luminous as a lighted torch, while they reflect a luftre on all neighbouring objects. They are in continual motion during the night; but the motion is various, and uncertain: fometimes they rife, and then fink. They will frequently difappear, and the next inftant rife in another place. They commonly hover about fix feet from the ground. It is faid, there is not a night in the year in which they are not feen. In the coldeft winter they are more frequently obferved, than in the warmeft fummer. Neither rain or fnow hinders their appearance. From all thefe circumftances

many

many fuppofe it to be the ignis fatuus, or the jack-in-the-lantern; which many have contended, is an inflammatory meteor, exhaled from marfhy lands, over which it is obferved to wander in the darkeft night.

GENUS V.

C I C A D A.

Character.

THE head bends downwards; the feelers are briftly; the four wings are membraneous; and the feet are adapted to leaping.

CICADA SPUMERIA.

The FOAMY FROG-HOPPER.

AMONGST the fpecies found in this country, of this genus, this is one of the largeft. It is a brown, tinged with green. The head, breaft, and elytra, are beautifully dotted: on the laft are two white fpots. Before the infect has metamorphofed itfelf, the larva which produces it, lives and refides upon plants: but it is not perceived, unlefs the fpot of its devouring is certainly known;

for by emitting, from every part of its body, foamy bubbles, refembling fpittlc, under which it conceals itfelf, the larva is not eafily difcovered : but when this froth is removed, the larva is found : but it is foon covered again, by a frefh emiffion of froth. Thus the larva is enabled by nature to preferve itfelf againft the injury of the weather, and from being deftroyed by other infects. This is another inftance of the variety of means adopted by the Creator to preferve the balance of all things. As the larva of this infect is liable to be preyed upon by different animals, it is provided with the power of emitting this foam, as the only protection againft its enemies.

CICADA SANGUINOLENTA.

The CRIMSON FROG-HOPPER.

THIS is thought the fineft fpecies which we, in this country, poffefs of this genus. The elytra alone have fix large beautiful crimfon fpots ; both the elytra are black at the extremity; and the wings are a dufky colour, and tinged with a little red at their bafe. This infect, not leaping much, is eafily taken ; but not near London; as it is very feldom found near the metropolis. It varies according to the different fize of the

T 2　　　　crimfon

crimfon fpots obferved on its elytra, or wing-
cafes.

─────────────

Genus XI.

C O C C U S.

Character.

THE trunk is placed in the breaft ; the
hinder part of the abdomen is briftly. The
males have two erect wings ; while the females are apterous, or without any.

─────────────

COCCUS PHALARIDIS.

The COCHINEAL FLY.

THE feet and body of this infect are nearly of a pink colour, and fprinkled with a
little white powder. The wings and four
threads of its tail, are of the cleareft white.
It is found on a fpecies of grafs called phalaris.
The female forms, on the ftock of this dog-
grafs, a white downy neft, in which fhe depofits her eggs. Being brought over with exotic
or foreign plants, they are fometimes found
in hot-houfes. This fpecies of gall-infect is
ufed in dying fcarlet. When the dried cochineal

chineal is fteeped in water, or vinegar, the parts of the body unfold themfelves; and become fo vifible, as to difplay even the ligaments of the legs.

The Indians in Mexico, where the propagation of the cochineal is a confiderable concern, gather them, and put ten or twelve in mofs, or the flue of the cocoa: they are then hung upon the thorns of the Indian figtree, which grows in great quantities round their habitations. They are fo prolific at to afford three gatherings of them every year. As foon as they are collected, they are deftroyed. Some they kill by the heat of ovens; and others by throwing them into hot water : while many are deftroyed upon the hot places ufed for roafting maize.— Three pounds of frefh cochineal weighs but one pound when dried. Cochineal will preferve, for ages, its colouring particles. This valuable infect is ufed for dying fcarlet and crimfon. The Indians mix it with gum lac, to dye their cloths. The cochineal furnifhes painters with many beautiful and fplendid tints. It is computed, that 880,000 lb. of thefe infects is imported yearly into the kingdom of Great-Britain. Where it propagated in the American iflands, where the climate is congenial with this infect, great advantages might be derived : and as the cochineals of Europe refemble greatly thofe of America, they might, probably, be productive of emolument.

THE

THE THIRD ORDER.

INSECTA LEPIDOPTERA.

LEPIDOPTEROUS infects have four wings, covered with fcales. The mouth has a fpiral tongue, which they unfold at pleafure. Their bodies are hairy.

This order is divided into three genera.

─────

GENUS I.

PAPILIO.

THE horns are thickeft at their extremities; and are in moft terminated by a kind of capitulum, or little head. When fitting, the wings are erected, and touch each other.

─────

GENUS II.

SPHINX.

Character.

THE horns are thickeft in the middle: refembling in form, a prifm. The wings are

are bent inwards. They are flow and heavy in their flights, which they take either late in the evening, or early in the morning.

Genus III.

P H A L E N A.

Character.

THE horns are briftly, decreafing in fize from the bafe to the point; which chiefly diftinguifhes it from the butterfly. The wings, when at reft, generally turn down. They fly in the night.

For a more particular defcription of butterflies and moths, fee our account, from page 179 to page 188.

THE

THE FOURTH ORDER.

INSECTA NEUROPTERA.

NEUROPTEROUS infects have four tranſparent, membraneous, and uncaſed wings, which are veined like net-work. Their tail is unarmed, or ſtingleſs; but it is frequently furniſhed with appendices, like pincers, by which the males are diſtinguiſhed.

LIBELLULA.—*The* DAMSEL.

THIS genus of infects is well known to every body. The largeſt ſpecies is produced from a water-worm, that has ſix feet, which yet young, and very ſmall, is transformed into a chryſallis, that has its dwelling in the water. People have thought they diſcovered them to have gills like fiſhes. It wears a maſk, as perfectly formed as thoſe that are worn at a maſquerade; and this maſk, faſtened to the infect's neck, and which it moves at will, ſerves it to hold its prey, while it devours it. The period of transformation being come, the chryſallis makes to the water-ſide, undertakes a voyage, in ſearch of a convenient place; fixes on a plant, or

ſticks

sticks fast to a bit of dry wood.   Its skin,
grown parched, splits at the upper part of
the thorax.   The winged insect issues forth
gradually, throws off its slough, expands
its wings, flutters, and then flies off with
gracefulness and eafe.   The elegance of its
slender shape, the richnefs of its colours, the
delicacy and refplendent texture of its wings,
afford infinite delight to the beholder.

In order to accomplish the purpofe of nature, the male, while hovering about, watch-
es, and then feizes the female by the head,
with the pincers with which the extremity
of its tail is armed.   The ravisher travels
thus through the air, till the female yeilds
to his fuperior ftrength.   Thefe flies are
feen thus coupled in the air, exhibiting the
form of a ring.   The female depofits her eggs
in the water, from whence fpring water-
worms, which afterwards undergo the fame
transformations.

LIBELLULA GRANDIS.

The GREAT DAMSEL.

THIS fpecies is the largeft of any this
country affords.   Its head is yellow, efpecially forwards; its eyes are brown, and being very large, meet on the top of the head,
and are often fet with dots, raifed and shining

ing, which would conftitute a very diftinc-
tive charaCter, if it were conftant; but fometimes thofe dots are abfent, or there are, at
moft, but one or two. The thorax is dun-
coloured, with two oblique bands on each
fide, of a lemon-colour. The abdomen,
which is very long, is likewife of a deep
buff, often fpotted with white on the top
and bottom of each fegment. The fmall
laminæ that terminate the abdomen are very
long in this fpecies. The wings have more
or lefs of the yellow dye, with a brown fpot
on the exterior edge. At the rife of each
wing there is a fmall protuberance, of a dark
brown colour.

====

LIBELLULA VIRGO.—*The* VIRGI N.

THIS beautiful libellula has a large head,
reticulated, prominent, brown eyes, that
are not in contaCt with each other. The
fpace intervening between the eyes, exhibits
the three brown ftemmata, placed in a trian-
gle. The neck, on which the head is refted,
is fhort and narrow. The thorax is larger,
of a bright green and blue colour. From
the inferior part of the thorax arife the fix
legs, long, and charged with a double row
of fmall fpines, a circumftance common to
this genus. From the upper part come forth
the four wings, all of equal fize. They are
                                        much

much reticulated, and have on their middle
a large cloud, of a blueifh brown, that oc-
cupies above one half of them. The bafe
and extremity of the wing are, the only parts
not charged with the fame colour, being only of a yellowifh hue. On the outer edge
of the wing there is no fpot; which is uncommon in this genus. The abdomen, long,
cylindric, and confifting of nine or ten fegments, is of a blue colour, fometimes bordering on green, and very bright. This
beautiful infect is met with in meadows, on
the banks of ponds.

LIBELLULA PUELLA.

THE wings of this infect are whitifh,
nicely veined with black, with a black fpot
on the exterior edge towards the extremity.
The colour of the head is a leaden blue,
with brown eyes. The thorax, which is
blue, is adorned with three brown longitudinal bands, one on the middle, and two narrower ones on the fides. The fegments of
the abdomen are blue, with a black ring
towards their pofterior extremity. They are
nine in number; the two laft larger than the
reft, and entirely brown. This infect is
found in meadows.

The remaining libellula is only a variety
in colour, the body being of a fine red.

U   GENUS

GENUS II.

EPHEMERA.

Character.

THE mouth has neither teeth nor spiral
tongue. The wings are erect, and the hinder shorteft. The tail is furnifhed with hairs,
or briftles. The horns are fhort and briftly.

EPHEMERA.—*The* DAY-FLY.

THESE flies derive their name from the
fhort period of their exiftence. Some of
their different fpecies live feveral days;
while others, that take their firft flight at the
fun, die before that luminary rifes again.
Some have only the life of an hour; others
exift but half an hour.

The ephemeræ, before they fly, have
been in fome manner fifhes: and, what is
very remarkable, they have been obferved
to remain as long as one, two, and three
years, in their larva and chryfalid ftates.
Both the larva and chryfalis have fmall
fringes of hair on each fide; which, when
moved in the water, ferve them as fins.
                                        The

The plying of thefe little oars is exceedingly curious. The larvæ make their refidence by perforating, or making holes in the banks of rivers; 'and, when the water falls, or decreafes, they make other holes lower, in order to have ready accefs to their favourite element. Flames attract them, fo as to caufe them to form a thoufand circles round fuch a light, with an amazing regularity. One fingle female will lay feven or eight eggs, which fink to the bottom of the water, where they are depofited. The larvæ which they produce, conftruct habitations to fhelter them from every danger. The flies, having propagated, immediately die in heaps.— Fifhermen confider thefe multitudes of deftroyed infects, as manna for the fifh. We can, therefore, perceive, that even this infect, which cannot, for its very fhort exiftence, be of much fervice during life, is, by the wifdom of the creator, fo calculated, as to be of effential fervice, even in its departed ftate.

Genus

Genus V.

MYRMELION.

Character.

THE mouth is armed with jaws, two teeth, and four long spiral tongues. The tail, in the male sex, is forked. Their feelers are club-formed, and as long as the breast : and the wings bent downwards.

═══════ .

MYRMELION.----The ANT-EATER.

AS few insects afford greater entertainment, or gratify curiosity more, by their wiles and stratagems, than this ; we shall forbear all uninteresting description, to confine ourselves to what we think more essential. Before the head of the larvæ, is placed a dentated forceps, with which they catch and suck flies, and ants especially. This animal having a retrograde motion, which prevents its being able to pursue its prey, it has recourse to the following stratagem. Having dived into the sand, or soft mould, it hollows out furrows, that meet in a centre, and grows deeper by degree. The superfluous sand it carefully removes from the scene of action ;

action ; after this, it digs a hole, like a funnel, at the bottom of which this animal stations itself, suffering only its extended forceps to be seen above it. Ruin awaits the insect that falls, unfortunately, into this cavity. The myrmelio, being apprised of its approach, by grains of sand rolling down to the bottom, immediately overwhelms the fallen prey with a shower of dust, which it casts with his horns. It then drags the poor captive to the bottom of the hole, where it is immediately destroyed. Such is the rapacity of this creature, that it will prey in this manner even on its own species. This is one of the few instances nature affords of any one sort of animal preying on its fellow-creatures. To the disgrace of man, this destruction of each other is very rarely sanctioned by example, in all the infinite course of being with which the creation abounds.

The perfect insect of the ant-eater is very seldom found ; when it is, it is chiefly in sandy places, near rivulets.

THE

THE FIFTH ORDER.

INSECTA HYMENOPTERA.

HYMENOPTEROUS insects have four membranous wings: and moft of their tails have ftings; except the males, which are harmlefs.

GENUS I.

CYNIPEDES.

Character.

THE mouth is armed with jaws; but has no trunk. The fting is fpiral, and concealed moftly in the body.

CYNIPS.---*The* GALL-FLY.

THIS infect is of a burnifhed fhining brown colour: the horns are black, and the feet chefnut; and the wings are white. The gall-fly is produced in thofe little fmooth, round, and hard galls, which are found faftened to the fibres under oak-leaves. This gall is caufed

by the overflowing of the fap of the leaf, occafioned by the fly having pierced it, for the purpofe of depofiting there its eggs.— Sometimes, inftead of the cynips, a large infect proceeds from the gall, and which is called an ichneumon. This latter infect is not the real inmate of the gall; he is a parafite, whofe mother depofited her egg in the yet tender gall; and, when hatched, produces a larva, that devours the larva found there of the cynips. Of this genus, there is a fpecies which caufes the galls of which the Norway ink is made.

Genus VIII.

A P I S.—*The* B E E.

THESE infects are divided into feveral fpecies, which are diftinguifhed from each other, by genius, talent, manner, and difpofition. Some live in fociety, and fhare the toils: others dwell, and work, in folitude, building the cradles of their families, as the leaf-cutter bee does, with a rofe leaf; the upholfterer, with the gaudy tapeftry of the corn-rofe; the mafon-bee, with plafter; and the wood-piercer, with faw-duft. But all, in general, are employed, in their little kingdom, with providing for their pofterity, and

and contributing to the general welfare of their community.

Of bees there are three forts; the plebeians, the drones, and the queen. The queen, or parent-bee, is the foul of the hive: to her all the reft are fo attached, that they will follow her wherever fhe goes. If fhe happens to die, all their labours are at an end, an univerfal mourning enfues, and all her fubjects die, by rejecting their food. Should a new queen arife, before this cataftrophe attends the hive, joy renovates their fpirits, and their toils are renewed. This has been tried by removing the chryfalis of a queen-bee from one hive to another, which had loft its own emprefs. But this attachment is only in proportion to the utility fhe affords to the commonwealth. She is fo prolific, that fhe lays 15 or 18,000 eggs, which produce 800 males, four or five queen bees, and the reft neuters, or plebeians. Their cells differ in fize; the largeft are for the males, the royal cells for the queens, and the fmalleft for the neuters. The parent-bee depofits in thofe cells fuch eggs as will produce the fpecies for which the refpective cells are deftined. In two or three days the eggs are hatched; when the neuters turn nurfes to the reft, which they feed moft tenderly, with unwrought wax and honey. After twenty-one days, the young bees are able to form colonies, with fuch indefatigable activity, that they will do

more,

more, in one week's time, than they will during all the reft of the year. Sometimes there are bees lefs laborious, who fupport themfelves by pillaging the reft of the hives; on which a battle enfues between the induftrious and the defpoiling infects. Frequently contentions will arife among them, when a new colony feek their habitation in a hive already occupied. Their foes are the wafp and hornet; which will rip open their bellies with their teeth, in order to fuck out the honey contained in the bladder. Sparrows, fometimes, are feen to take one in their bill, and one in each of their claws.

The neuter bees collect from flowers their honey and unwrought wax : they roll themfelves over the ftamina, and thus caufe the dufty effence to ftick to the hairs which cover different parts of their bodies. Being thus laden, they proceed with their burden to the hive; where they are met by other bees, that fwallow the wax they bring; this being afterwards refined in the laboratory of their ftomachs, is again produced by the mouth, as genuine wax, in the form of dough, which is next moulded into cakes of an admirable ftructure.

From the nectarious effluvia of flowers, the bee collects the honey, by means of its probofcis, or trunk; which is a moft aftonifhing piece of mechanifm, confifting of more than twenty parts. Entering the hive, the infect difgorges the honey into cells, for winter

ter fubfiftence ; or elfe prefents it to the la-
bouring bees. A bee can collect, in one
day, more honey than a hundred chemifts
could extract in a hundred years.

When they begin to form their hive, they
divide into four parties : one is deputed to
the fields, to collect materials; another is
ordered to work on thefe materials ; a third
is left to polifh the rough work of the cells,
and a fourth is allotted to provide for the la-
bourers. There are waiters always attending, to ferve the artizan with immediate re-
frefhments, left he fhould be too long abfent
from his work, by going to gather it himfelf.

So expert are thefe bees, that an honey-
comb, compofed of a double range of cells
backed one againft another, and which is a
foot long, and fix inches broad, is completed
in one day, fo as to contain 3000 bees. The
cells are moft curioufly compofed of little
triangular fides, which unite in one point,
and exactly conform to the like extremities
of the oppofite cells, refpectively. At every cell, the Creator has, moft wifely, taught
them to form a ledge, which fortifies each
aperture againft the injuries they might receive from the frequent ingrefs and return of
the bees.

How grateful ought we to be for the creation of this admirable infect ! To his toil
and wifdom we are indebted for one of the
moft agreeable and wholefome fubftances afforded,

forded by nature. Were it not for the bee, thefe flowery fweets would be loft in " the " defert air," or deeline with the fading flower. All the various ufes to which wax is applied, would be loft to man, had not the bee an exiftence.

Genus IV.

Character.

THE mouth has jaws, without any tongue. The horns eontain more than thirty joints; and the abdomen is generally joined to the body by a pedicle. The fting is inclofed in a eylindrieal fheath, eompofed of two valves.

The I C H N E U M O N.

ONE diftinguifhing and ftriking charaƈter of thefe fpecies of flies is, the almoft continual agitation of their antennæ. The name of Ichneumon has been applied to them, from the fervice they do us, by deftroying eatezpillars, plant-lice, and other infeƈts ; as the Ichneumon and mangoufte deftroy the erocodile. The variety to be found in the fpeeies of Ichneumons is prodigious among the

fmaller

fmaller fpecies. The males perform their courtfhips in the moft paffionate and gallant manner. The pofterior part of the females. is armed with a wimble, vifible in fome fpecies, no ways difcoverable in others : and that inftrument, though fo fine, is able to penetrate through mortar and plafter. The ftructure of it is more eafily feen in the long-wimbled fly. The food of the family to be produced by this fly, is the larva of wafps, or mafon-bees; for it no fooner perceives one of thofe nefts, than it fixes on it with its wimble, and bores through the mortar of which it is built. The wimble itfelf, of an admirable ftructure, confifts of three pieces : two collateral ones, hollowed out into a gutter, ferve as a fheath; and contain a compact, folid, and dentated ftem; along which runs a groove, that conveys the egg from the animal, which fupports the wimble with its hinder legs, left it fhould break ; and, by a variety of movements, which it dextroufly performs, it bores through the building, and depofits one or more eggs, according to the fize of the Ichneumon, though the largeft drop but one or two. Some agglutinate their eggs upon caterpillars eggs, though very hard, and depofit their own in the infide : when the larva is hatched, its head is fo fituated that it pierces the caterpillar, and penetrates to its very entrails : thefe larvæ pump out the nutritious juices of the caterpillar, without attacking the vitals of the

creature ;

creature; which appears healthy, and even
fometimes transforms itfelf to a chryfali .
It is not uncommon to fee caterpillars fixed
upon trees, as if they were fitting upon
their eggs; and it is afterwards difcovered
that the larvæ, which were within their bodies, have fpun their threads, with which,
as with cords, the caterpillars are faftened
down, and fo perifh miferably.

The ichneumons performed fpecial fervice
in the years 1731 and 1732, by multiplying
in the fame proportion as did the caterpillars:
their larvæ deftroyed more of them than
could be effected by human induftry. Thofe
larvæ, when on the point of turning into
chryfalids, fpin a filky cod. Nothing is
more furprifing and fingular, than to fee
thofe cods leap, when placed on the table, or
hand. Plant-lice, the larvæ of the curcu-
liones, fpider's eggs, are alfo fometimes the
cradle of the ichneumon-fly. Carcafes of
plant-lice, void of motion, are often found
on rofe-tree leaves. They are the habitation of a fmall larva; which, after having
eaten up the entrails, deftroys the fprings and
inward economy of the plant-loufe, performs
its metamorphofis under fhelter of the pel-
licule which enfolded it, contrives itfelf a
fmall circular outlet, and fallies forth into
the open air.

There are ichneumons in the woods,
which dare attack fpiders, run them through
with their fting, tear them to pieces, and

X   thus

thus avenge the whole nation of flies of fo formidable a foe: others, deftitute of wings (and thofe are females), depofit their eggs in fpiders nefts. The ichneumon of the bedeguar, or fweet-briar fponge, and that of the rofe-tree, perhaps, only depofit their eggs in thofe places, becaufe they find other infects on which they feed.

The genus of the ichneumon flies, might, with propriety, be termed a race of diminutive canibals.

Genus IX.

FORMICA.

Character.

A LITTLE upright fcale is fituated between the breaft and the belly. The feelers are broken, and have the firft articulation longer than the reft. The females and neuters have a fting, concealed in the abdomen. The males and females are winged; and the neuters are apterous, or without wings.

FORMICA.—*The* ANT.

NOT to impofe upon our readers thofe fables which have been related of this remarkable infect, we fhall confine ourfelves to the moft authentic accounts, and to our own obfervations in what we fhall briefly mention refpecting the ant. Sanctorius fays, when the ants carry any corn to their habitations, they carry it, exactly in form and intention, as they do, bits of wood, for the conftruction of their dwellings merely. For what purpofe fhould they provide corn for the winter, when they pafs that feafon without motion? But, from what we have lately obferved ourfelves, we rather imagine this error arofe from fome perfons having feen them dragging a number of their aurelias, when they have been removed, by a hoe or fpade, again to their repofitories; for thefe aurelias are exactly of the fize and colour of a grain of wheat. The great prudence ants difcover, is in fheltering themfelves from cold, which, when fevere, almoft deprives them of motion.

At the beginning of March, if the weather be warm, they go abroad in fearch of nourifhment. If corn be thrown to ants, they remove it from place to place, by fome dragging, others lifting, and two or three more pufhing forward, the weighty maffes.

A grain

A grain of wheat muſt be conſidered in proportion to their ſize and ſtrength. They have the precaution to make a bank near ſix inches high, above the entrance; and to make ſeveral roads, to go out and in, by what may be called their terrace-walk. From May or June, they work until the ſeaſon's change diſcontinues their induſtry. This labour is entirely for the preſervation of their brood, which is produced during the fine weather. When they attack fruit, they tear it into ſmall bits, and thus is each ant enabled to carry home his provender. Liquors which are ſweet, they have a mode of ſaving and carrying ſome for their young. They ſend their foragers to ſeek for food: if one of them proves ſucceſsful in finding ſome, he returns to inform the republic, and immediately ſallies from the town, to capture the prize. To prevent any delay, obſtruction, or confuſion, they have two tracks; one for the party loaded, and the other for that which are going to load themſelves. Should any be killed, ſome of them inſtantly remove the ſlain, to a diſtance. When proviſions are ſcarce, they portion them according to their preſent and future wants.

A neſt of ants is a ſmall well-regulated republic, united by peace, unanimity, good underſtanding, and mutual aſſiſtance. Great police in their little labours, prevents among them thoſe diſorders which frequently embarraſs and perplex the happineſs of even man,

man, who affumes to himfelf the title and
confequence of Lord of the creation. Each
ant has its talk affigned it; whilft one removes a particle of mould, another is re-
turning home to work. They never think
of eating, until all their talk is performed.
Within their common, but fubterraneous
hall, which is about a foot deep, they affemble, from their focial communities, fhelter
themfelves from bad weather, depofit their
eggs, and preferve their aurelias ; which, refembling grains of corn, as was obferved before, has caufed many to miftake them for
their granaries.

X 2   THE

THE SEVENTH ORDER.

INSECTA APTERA.

APTEROUS infects are distinguished from those of every other order, by neither sex having wings.

Species 1, Is a small spider of a scarlet colour. They are found in woods, and likewise on trees in gardens. They are the only species of spiders that are thought to be venemous, except the tarantula: for spiders are, in general, more frightful than injurious.

Species 2—Has six eyes. The colour is chiefly dark, with a broad streak of light colour in the middle of its back; and the form of a diamond, of the same colour, on the upper part of its belly. The legs are beautifully spotted.

Species 3.—This small long-legged spider is so finely marked, that it is impossible to describe it, either in words or colours; there being so admirable a combination of green, red, and black, interchangeably disposed into the most agreeable forms. The legs are curiously marked with the same colours. Its small eyes are not discernable.

Species 4.—This is one of the leaping spiders. It has eight eyes, placed in a circle; and all that have their eyes thus disposed

sed

fed, leap at their prey, like a cat feizing a moufe. It is extremely nimble. When viewed through a microfcope, its beauty appears unparallelled. Black, chefnut, red, and white, are moft admirably difpofed into the moft beautiful forms; but to the naked eye, it only appears rough, hairy, and grey-fpeckled. Dr. Hook gives the following diverting account of this fpider, as defcribed by Mr. Evelyn in his travels through Italy.

" Of all forts of infects," fays he, " there
" is none has afforded me more diverfion than
" the fmall grey jumping fpider, prettily be-
" fpecked with black fpots all over the body,
" which the microfcope difcovers to be a
" kind of feathers, like thofe on butterflies
" wings, or the body of the white moth.
" It is very nimble by fits, fometimes run-
" ning, and fometimes leaping like a grafs-
" hopper; then ftanding ftill, and fetting
" itfelf on its hinder legs, will very nimbly
" turn its body, and look round itfelf every
" way. Such," fays Mr. Evelyn, " I did
" frequently obferve at Rome, which, efpy-
" ing a fly at three or four yards diftance,
" upon the balcony where I ftood, would
" not make directly to her, but crawl under
" the rail, till, being arrived right under
" her, it would fteal up, feldom miffing its
" aim; but, if it chanced to want any thing
" of being perfectly oppofite, would, at
" the firft peep, immediately flide down
" again;

" again ; till, taking better notice, it would
" come, the next time, exactly upon the
" fly's back ; but, if this happened not to
" be within a competent leap, then would
" this insect move so softly, as the very
" shadow of the dial seemed not to be
" more imperceptible, unless the fly moved ;
" and then would the spider move also in the
" same proportion, keeping that just time
" with her motion, as if the same soul had
" animated both those little bodies ; and,
" whether it were forwards, backwards, or
" to either side, without at all turning her
" body, like a well-managed horse : but if
" the capricious fly took wing, and pitched
" upon another place, behind our huntress,
" then would the spider whirl its body so
" nimbly about, as nothing could be ima-
" gined more swift ; by which means, she
" always kept the head towards her prey,
" though, to appearance, as immoveable as
" if it had been a nail driven into the wood,
" till, by that indiscernible progress, being
" arrived within the sphere of her reach,
" she made a fatal leap, swift as lightning,
" upon the fly, catching him in the pole,
" where she never quitted hold until her
" belly was full, and then carried the re-
" mainder home. I have beheld them in-
" structing their young how to hunt ;—
" which they would sometimes discipline
" for not well observing ; but when any of
" the old ones did miss a leap, they would

" run

" run out of the field, and hide themfelves
" in their crannies, as afhamed, and not be
" feen abroad for four or five hours after;
" for, fo long have I watched the nature of
" this ftrange infect, the contemplation of
" whofe wonderful fagacity has amazed
" me : nor do I find, in any chace whatfo-
" ever, more cunning and ftratagem ob-
" ferved. I have found fome of thefe fpi-
" ders in my garden, when the weather,
" towards the fpring, is very hot; but they
" are nothing fo eager of hunting as they
" are in Italy."

Species 5.—This is called the carter, or
long-legged fpider. It has only two eyes,
which are moft curioufly placed on the top
of a fmall pillar, rifing out of the top of the
back. The eyes have a black purple in the
centre of the cornea, and the iris of them
is grey. It is likewife remarkable for the
length of its legs, and diminutive body.
The legs are alfo jointed like thofe of a
crab; and each terminates in a fmall fhell
cafe, fhaped like that of a mufcle : they are
faftened to the body, in a manner that moft
curioufly difplays the wonderful mechanifm
of nature. Thus is the infect enabled to
move, with the greateft celerity, over the
tops of grafs and leaves, where it fearches
for its prey. The head, breaft, and belly of
this creature, are fo indifcriminated by nature, that it is fcarcely poffible to difcern
the one from the other. Many fuppofe it

to

to be meant by the Creator as the air crab;
and adapted to the light element, in the
fame proportion as the fea crab is adapted
for the water.

Genus VIII.

A R A N E A.

Character.

THIS infect has eight feet, as many eyes,
a mouth armed with two crotchets, two
fpiral tongues; and the bottom of the abdomen has two inftruments, like nipples, adapted for fpinning.

Of thefe infects there are many different
fpecies. That which moftly diftinguifhes
the fpider, is the manner of forming its
web: fhe firft choofes a place where there is
a cavity, that fhe may have a clear paffage,
to pafs freely on each fide, and to efcape occafionally. She begins, by dropping on the
wall fome of her gum; to which fhe attach-
es her firft thread, which lengthens as fhe
paffes to the other fide, to which fhe fixes
the thread in a fimilar manner: thus fhe
paffes and repaffes, from fide to fide, until
fhe has made what may be termed the warp
of her web, exactly the fize fhe intends it
fhould be, or which fhe thinks will anfwer
her

her purpofe of preying on the paffing fly.
It is obferved that in order to finifh her work
the fooner, fhe fpins feveral threads at one
time : after thus finifhing, fhe then croffes
her work with threads, in the fame direction
as the weaver throws the woof with his fhut-
tle.   To prevent her being feen, fhe weaves
a fmall cell in the web, where fhe lies,   un-
obferved, until the tremulous thread informs .
her of fome prey being entangled in her
toils : fhe then darts along the line, and
feizes the victim, then devoted to deftruction.
Many fuperficial obfervers of nature have
wondered from whence the fpider could be
fupplied with the gum fhe ufes in the many
webs fhe is obliged to make, or repair : they
never reflected, that the fame providence
which knows the fpider is hated, and that
her web is always in danger of injury,
could furnifh her with a magazine of both
gum and thread, for fuch exigencies ; and
that when the magazine was exhaufted, it
could, by the fame means, be replenifhed.
However, it muft be admitted the recruits
fail in time ; for when the infect grows
old, it is deprived of its weaving materials : it is therefore obliged to depend on the
generous compaffion of the young fpider,
who will frequently refign its own web to
the infirm infect, and weave for itfelf another.
   The web of the garden fpider differs almoft as much from the web of a houfe fpider,

der, as a net does from a clofe-weaved piece
of cloth : but it is, perhaps, more curious
in its formation.   They greatly refemble a
wheel, that has bars croffing the fpokes at
equal diftances.   Thefe fpaces are in proportion to the fize of the prey the fpider defigns
fhall not pafs through them. Being too fmall
for large flies, moths, butterflies, &c. to
pafs through with their expanded wings,
fuch generally fall the victims of the fpider,
whenever they unknowingly fly againft its
web.

Having given this general defcription of
what is moft extraordinary in the fpider, we
fhall now fay a few words on the

ARANEA DIADEMA.

The DIADEM'D SPIDER.

THIS infect grows very large. The upper
part of its belly is moft beautifully embellifhed with black and white dots and circles ;
in the middle of them is a band, compofed
of oblong fhaped fpots, of a pearl colour ;
refembling, in their arrangement, the fillet
of an eaftern king : the ground of this fillet,
when viewed in the fun, through a glafs, is
perhaps one of the richeft and moft fplendid
fpectacles nature has to exhibit, in all her
tribe of infects. The eyes are eight in
number,

number, sparkling and placed on the crown of the head : the legs are long, yellow, encircled with dark brown, and furnished with briftles.

The TARANTULA.

THIS insect being of this genus, and much resembling a houfe spider, we shall clofe our brief system of infects, with a few words on this extraordinary animal. The bite of it, in hot countries, producing the mott astonishing effects, naturally first arrests our attention. The quantity of the poifon emitted into the wound, is too inconfiderable to render it immediately perceptible; but, as it ferments, it caufes, in about five or six months, the moft frightful diforders. The perfon bit, at this time laughs and dances inceffantly, is all agitation, and affumes a moft extravagant fpecies of gaiety ; or clfe is afflicted with a moft difmal melancholy. At the return of the period when the bite was given, the madnefs renews ; and the diftempered party repeats his former inconfiftencies, by fancying himfelf a king, or a fhepherd, or fome other character, according as his fhipwrecked reafon is driven againft the rocks of abfurdity. He has no regular train of thought ; all his mind and feelings are but a chaos of wildnefs and extrava-

Y   gance,

gance. Sometimes these unhappy symptoms
will continue several years, until death relieves the sufferer.   Those who have been
in Italy, where the natives are frequently
afflicted with this malady, tell us, the only
cure is music, from such an agreeable and
sprightly instrument as the violin, which is,
therefore, one of the most common species
of music in that country; no village, or
cottage, is scarcely without it.   The tune is
chosen according to the natural temper and
disposition of the patient: this is discovered
by playing several tunes, until the unhappy
sufferer, by his gestures, shows that one is
found agreeable to his fancy: this is thought
an infallible sign of a cure being effected.
The patient immediately begins to dance,
and rises and falls in concert with the modu-
lations of the tune.   This is continued until
he begins to perspire, which instantly causes
an external evacuation of the venom.   In
this manner are those afflicted with the bite
of a tarantula, cured.   But, is it not an extraordinary instance of providence, that in-
strumental music should have attained so
great and general a perfection as it has in
Italy, where it is necessary to preserve the
lives of the natives, who would otherwise
frequently die from the bite of this baneful
and venemous insect?

The

The Z I M B.

HAVING obferved a curious account of the zimb, in the travels of Mr Bruce, we could not refrain from extracting it, as a moft valuable addition to our fmall compendium of natural hiftory.

This infect is called the zimb, or tzalfalya. It is a little larger than a bee; with wings of pure gauze. The head is large; the upper jaw fharp, and furnifhed with a fharp-pointed hair, about a quarter of an inch long : the lower jaw has two of thefe pointed hairs; and the three, joined into one pencil, make a refiftance to the finger, nearly equal to that of a hog's briftle. As foon as this winged affaffin appears, and his buzzing is heard, the cattle forfake their food, and run wildly about the plain, till they die, worn out with fatigue, affright, and pain. The inhabitants of Melinda, down to Cape-Gardefan, to Saba, and the fouth coaft of the Red Sea, are obliged to put themfelves in motion, and remove to the next fand, in the beginning of the rainy feafon: this is not a partial emigration; the inhabitants of all the countries, from the mountains of Abyffinia, northward, to the confluence of the Nile, and Aftaboras, are once in a year, obliged to change their abode, and feek protection in the fands of Beja.

The

The elephant and rhinoceros, which, by reafon of their enormous bulk, and the vaſt quantity of food and water they daily need, cannot ſhift to defert and dry places, are obliged, in order to refiſt the zimb, to roll themſelves in mud and mire, which, when dry, coats them over like armour. ·

Of all thoſe who have written of theſe countries, the prophet Iſaiah alone has given an account of the zimb, or fly, and deſcribed the mode of its operation. Iſaiah, chap. vii. ver. 18 and 19. Providence, from the beginning, it would appear, had fixed its habitation to one ſpecies of foil; which is a black, fat earth, extremely fruitful. And, contemptible as it ſeems, this infeẟ has invariably given law to the ſettlement of the country : it prohibited, abſolutely, thoſe inhabitants of the black earth, called Mazaga, houſed in caves and mountains, from enjoying the help of labour of any beaſts of burden. It deprived them of their fleſh, and milk, for food; and gave riſe to another nation, leading a wandering life, and preſerving immenſe herds, by conduẟing them into the ſands, beyond the limits of the black earth, and bringing them back when the danger from this infeẟ was over.

In the plagues brought on Pharaoh, it was by means of this infeẟ that God ſaid he would ſeparate his people from the Egyptians. The land of Goſhen, the poſſeſſion of the Iſraelites, was a land of pailure, not tilled,

tilled, nor fown, becaufe not overflowed by the Nile; but the land overflowed by the Nile, was the black earth of the valley of Egypt: and it was here that God confined the zimb; for he fays, it fhall be a fign of this feparation of the people, which he had then made, that not one fly fhould be feen in the fand, or pafture-ground, the land of Gofhen. And this kind of foil has ever fince been the refuge of all the cattle emigrating from the black earth, to the lower part of Albara: fo powerful is the weakeft inftrument, in the hands of the Almighty.

Y 2
A CON-

A

CONCISE DESCRIPTION

OF THE

MOST VALUABLE AND CURIOUS

TREES, SHRUBS, &c.

●━◆━◆━◆━◆━◆━●━●━●━●━◆━§━◆━●━●━◆━◆━◆━━◆━◆━◆━●

A CONCISE

DESCRIPTION, &c.

————————————

IN this part of our natural hiftory, which we have devoted to the fubject of trees, we have felected thofe of foreign production with which we are moft interefted, from their being the firft objects of our commerce, and the moft valuable of our exotic delicacies. Under this head of trees, we mean to treat of fuch plants and fhrubs as are particularly deferving the attention of our young ftudents, whether defigned for the fenate, clofet, counting-houfe, or counter.

═══════════════

COFFEE SHRUB.

THE coffee fhrub grows in Arabia-Felix, and is brought from Mocha : the flower refembles the jeffamine ; and the leaf, that of the bay-tree. It is propagated by feeds, and

grows

grows to the height of eight or ten feet.
The twigs and leaves rife by pairs : the
leaves are two inches broad in the middle,
from whence they decreafe to a point at
each extremity.   As this tree will not thrive
when tranfplanted, unlefs kept in mould, it
has been found very difficult to rear it in
diftant climates : but this inconvenience has,
by attention and perfeverance, been fo confiderably diminifhed, that it is now cultivated, with the moft promifing fuccefs, in the
Weft as well as the Eaft Indies.

The fruit hangs on the twigs, by a footftalk, containing one, two, or more, in the
fame place.   Thefe fhrubs are watered by
artificial channels, like other vegetables ;
and, after three or four years bearing, the
natives plant new fhrubs, in confequence of
the old beginning then to decline.   They dry
the berry in the fun, and afterwards diveft
it of the outward hufks, with hand-mills.   In
the hot feafons, they ufe thefe hufks, roaft-
ed, inftead of the coffee berries ; and efteem
the liquor impregnated with them   more
cooling.

The coffee berries are generally ripe in
April : they are efteemed, as being of an
excellent drying quality, comforting the
brain, eafing pains in the head, fuppreffing
vapours, drying up crudities, preventing
drowfinefs, and reviving the fpirits.

TEA

TEA SHRUB.

THE tea fhrub grows plentifully in feveral parts of the Eaft-Indies, and affords a leaf
which is too well known, according to the
opinion of our phyficians, in every country
in Europe. It is brought from China, Japan, and Siam. The leaves are gathered in
the fpring ; and bear a flower of five leaves,
refembling a rofe ; to thefe fucceed a cod,
like a hazle-nut. The tea fhrub flourifhes
equally in rich and poor ground. The leaves
are dried and parched by fire ; in which
ftate they are fent to Europe, and other
parts of the world. The beft tea is that
which is the greeneft, beft fcented, and moft
free from duft. The caufe of tea being fo
much drunk in Europe, is faid to be from
the Chinefe bartering it for their fage, which
they efteem as poffeffing the moft invaluable
qualities. This is not improbable, from our
phyficians having a Latin proverb, refpecting
fage of virtue ; which afks, why will a man
die with fage in his garden ? Although tea
is drunk more for pleafure than for any medicinal purpofe, it is juftly allowed to poffefs
many falutary qualities.

COCOA-TREE.

THIS tree, bearing the cocoa or chocolate nut, resembles our heart cherry tree; except that, when full grown, it is much higher and broader. It has abundance of leaves, similar to those of the orange-tree. It flourishes throughout the year, especially near the summer and winter solstices. As the leaves perpetually replenish themselves, this tree is never disrobed of its verdure. The blossoms are small, regular, and like a rose, but scentless. Every blossom is joined to the tree by a slender stalk; and leaves, in falling, long green filaments; which produce a pointed yellow fruit, of the size of our melons: these adhere to the thick branches, without any intermediate stem; as if nature thus providentially provided it a support strong enough to bear the greatness of its weight, when grown ripe, and to its largest size. Each fruit contains from between 15 and 25 small nuts, or almonds, covered with a thin yellow skin; which being separated, a tender substance appears, divided into several unequal particles, that, although sharp to the palate, are nourishing to the constitution.

These trees grow in all the Spanish West-Indies, Jamaica, &c. where they commonly produce fruit every seven years at most,

after

after the firſt planting : but, in the interim,
they are fometimes twice or three times removed ; when great care is taken to fecure
them, with fuch fhade as may preferve them
from the intenfe heat of the fun. Being
once reared, they are not liable to this injury : and, therefore, the precaution being no
longer neceſſary, is difcontinued; for, being
ranged in rows, with fhady plantains, they
are both mutually fheltered by each other
from the parching fun, and boifterous winds.
It is a tree of fingular beauty, profit, and
utility. , Its large, broad, and green leaves,
hang like fo many fhields, as if to defend
the tender and valuable fruit from injury.
As the fruit adheres to the large branches,
the tree appears as if moſt beautifully ſtud-
ded, from the root to the moſt large and expanding branches.

The cocoa-nuts, affording to the Indians
and Spaniards food, raiment, riches, and
delight, are received in payment, as cur-
rency.

It is unneceſſary to add, that, from this
extraordinary tree, that wholefome beve-
rage chocolate is made, in fuch quantities as
to fupply the greater part of the world with
a liquor diftinguifhed for its nutritive and reftorative qualities.

Z.   The

The SUGAR-CANE

IS the produce of Barbadoes, Jamaica,
Nevis, &c. This plant bears on each joint
a cane, five or six feet high, and adorned
with long, strait, green leaves, similar to
flags, or fleur-de-lis. On the top they have
a plume of silver-coloured flowers. The
canes contain a porous substance, of which
the sugar is made. When they are mature,
the canes are cut off, at the first joint from
the ground; and are laid in heaps, like our
sheaves of corn in harvest-time: being cleared
from their leaves, they are tied in bundles,
and carried to the mills, which press out their
juice: this is put into boilers, in order to
evaporate the watery particles, so as to let
nothing but the sugar subside. The sugar is
then cleared, by a mixture of ingredients,
adapted to the purpose of fining and prepar-
ing it for graining. While it is boiling, the
scum, which rises in great quantities, is clear-
ly taken from the surface, until the sugar is
ready to be emptied in the coolers; from
whence it is again shifted into earthen pots,
with holes in their bottoms, which drain
the molasses into other pots, placed beneath:
the latter is an entire month in separating itself from the sugar; which is then put into
casks, or hogsheads for transportation. The
sugar-cane, in England, is so tender as not

to

to admit of being reared without artificial heat. It is, however, preferred as a great curiofity, in the gardens of thofe who keep hot-houfes, for the purpofe of having fuch curious and exotic productions of nature.

The NUTMEG and MACE TREES.

NUTMEGS are diftinguifhed by the fexual difference of male and female ; but the latter is the moft ufeful, and therefore moft valuable. The male is long, and large ; the female is round, and fmall, which only grows in improved or cultivated lands : while the males, growing fpontaneoufly in woods and forefts, are called by the Dutch, the wild nutmegs. The tree which produces the female, or beft nutmeg, is as large as a pear-tree, and has leaves fhaped like thofe of the peach. The bloffom has a pleafant odour, and refembles the rofe. The flower being fallen, a fruit appears, as large as a green walnut : in this is a kernel, which is the nutmeg. It has two barks : the firft is very thick, and is taken off when the fruit is ripened ; the other is thin, and of a reddifh yellow. When feparated from the nutmeg, it is dried, and called *mace*. The nutmegs being divefted of their bark, are dried and preferved.

The nutmeg-trees grow plentifully in the Afiatic Ifland of Banda, and in feveral other iflands in that part of the Eaft-Indies which belongs to the Dutch, who are the fole poffeffors of this produce. It is faid thofe iflands fo abound with nutmeg-trees, as would appear incredible to relate: and the climate is fo fertile, and fo congenial to their nature, that they produce three crops annually, in the months of April, Auguft, and December.

According to Tavernier, this tree is not planted, but grows by means of certain birds, which fwallow the fruit whole, and afterwards void it, in its perfect ftate, but covered with a vifcous or gluey matter. Being thus prepared for vegetation, they take root wherever they fall, and produce the trees above mentioned.

CINNAMON-TREE.

THIS tree affords a bark, which is the cinnamon, fo well known as one of the moft valuable of the fpices confumed in Europe. The tree itfelf is about the height of the willow: it bears little blue cups, which are odorous; and are fucceeded by the fruit, refembling the olive.

This tree grows fpontaneoufly in the ifland of Ceylon, which is poffeffed by the Dutch.

There

There are nine or ten forts of cinnamon :
the beft grows in the greateft plenty, and is
the peculiar produce of that ifland. The
natives call it *raffe corronde*, i. e. fharp,
fweet cinnamon. The Dutch Eaft-India company export it annually, under the ftricteft
orders of no other cinnamon being mixed
with it. Every fort of cinnamon-tree muft
grow a certain number of years before it is
ftripped of the bark. Thofe growing in
vallies, of a white fandy foil, will ripen in
five years ; while others, found in a wet,
flimy foil, will be at leaft feven or eight
years before they can be ftripped : and fuch
as grow in the fhade of larger trees, are
not only later, but produce a bark not fo
fweet or agreeable as the more early cinna-
mon-trees. The bad cinnamon taftes bitter,
and fmells like camphire. The fweetnefs is
entirely owing to a thin membrane, which
adheres to the infide of the bark. The flavour diffufes itfelf through the whole fubftance, while the cinnamon is drying in the
fun. The fragrancy of the fmell, and the
fweetnefs of the tafte, have caufed this
fpice to be coveted by all nations. The
bark may remain on fome trees, 14, 15, or
16 years, without fuffering any material di-
minution in its qualities : but after this period, the tafte and fmell decreafe, and approach to thofe of camphire. The cinnamon ftripped from trees that are too aged,
may be known by its being thick, and con-

fequently flat; from the fun not having the
power of warping it in the drying. The
amazing quantities imported into Europe,
and other parts of the world, are falfely faid
to be produced by the trees barking again,
in four or five years: the real caufe is, that
the trees, being cut down to the ground,
fprout branches, which grow, and ripen, fo
as to produce bark in five, fix, feven, or
eight years. A fpecies of dove, likewife,
contributes greatly to the confiderable produce of cinnamon. Thefe doves are called
cinnamon-eaters, from eating vaft quantities,
and difperfing its fruit over the fields, for
the fubfiftence of their young. Thus is the
vegetation of the cinnamon-tree extended
over the whole ifland.

The oil drawn by fire from cinnamon, is
efteemed as one of our firft cordials. The
camphire, which is extracted from the root,
is a moft ufeful and valuable medicine. Oil
of camphire is very coftly; not fo much
from its fcarcity, as from its medicinal effi-
cacy. In a word, there is no part of the
cinnamon-tree but is ufeful.

CLOVE-TREE.

THIS tree produces a flower, the foot-ftalk
of which is what we call cloves. The fruit,
when ripe, is a dark brown. The trees
grew

grew moſt plentifully in the Molucca Iſlands, until the Dutch pulled them up, to prevent the produce being ſhared by the Engliſh, and other nations. They were then tranſplanted to an iſland called Ternati, which was in the entire poſſeſſion of the Dutch. Thus every other people is obliged to purchaſe from them this valuable merchandiſe.

The cloves are only pulled from the trees, ſpread in the open fields, and thus dried in the ſun : the only care that is afterwards required, is to preſerve them from the air. Some authors deſcribe the royal clove, ſo called from bearing on its top a crown ; which is one reaſon of the king of this iſland keeping it in his poſſeſſion ; and from the fabulous opinion, that the other trees bow to this, as their ſovereign.

PEPPER-TREE.

THE fruit of this tree is the black Eaſt-India pepper : it grows in the manner of a climbing vine or creeper, and produces the fruit in ſmall cluſters like our currants. The ripe ſeeds are about the ſize of a large currant, which turns, in drying from a red to a black colour. It is ſaid the common white pepper is only the black ſtripped of its outward ſkin, which is effected by ſteep-ing it in ſea-water, then drying and rubbing

it

it in the fand. There is, however, a natural white pepper poffeffing all the qualities of the black. Three forts of black pepper are brought from the Eaft-Indies by the Englifh and Dutch, which only differ in the places from whence they are brought : the fineft comes from Malabar. The tree or bufh bearing the Jamaica pepper grows nearly like the Barberry, except not being fo high, and having no prickles. The berries refemble thofe of the juniper, poffefs an aromatic tafte, which, partaking of thofe of all other fpices, has caufed it to be called *all-fpice*. This pepper grows plentifully in many of the plantations in Jamaica.

GINGER PLANT.

THIS plant is called the club-reed ; from the root of which is the ginger, which, at the end of every root, is in form like a foot. The leaves of the plant are long, large, and of a deep green : and the whole flower refembling a club, has caufed it to be called by fome the *club-reed*, and by others *ginger with a club flower*. Ginger confifts of one fort which is white and mealy, and another which is black and hard; the firft is the moft efteemed. Both the Eaft and Weft Indies produce ginger : in the Antilles it is greatly cultivated : but the greateft quantities are

imported

imported from the leeward island· of Barbadoes, Nevis, St. Chriftopher, and Jamaica. Little is now brought from the Eaft- ndies, except what comes as confectionary, and is called green ginger, which they prepare in India. Some indeed is prepared in England and other parts, by fteeping the frefh roots two or three days in warm water, keeping it all the time in a balneo, which fwells and foftens it. It is then boiled, either flit or whole, with refined fugar, until it becomes a fyrup.

CURRANT VINES.

THIS vine grows moft plentifully in a fpacious plain near the fortrefs of Zant in Greece. It produces thofe currantswhich are called the *Corinth grape*, vulgarly *currants*, and are fold by our grocers for cakes and puddings. They confift of three forts, the red, black, and tawny. The vine itfelf is low, has thick indented leaves and is furnifhed like other vines with clafpers at the joints. Thefe little grapes, which grow in bunches, ripen in Auguft, when the people of Zant gather, ftone, and dry them. They are then carried into the town, and depofited through a hole, in the grand magazine called the Seraglio, where they are preffed in fo compact a mafs, that it is obliged to be cut with an iron inftrument, in order to pack
them

them in cafks and bales for exportation.—
Thefe currants are likewife brought from
feveral parts of the Levant; but the fort
we moftly ufe, comes from the iflands near
the Morea. The people near Zant fuppofe
we ufe them in dying inftead of eating. The
raifins fold alfo by our grocers are grapes
from vines growing in this country, and
which are dried and packed in a fimilar
manner to the currants, but with the difference of their not being ftoned.   Some indeed affert that, before they expofe thefe
vine branches to dry in the fun, they are
firft dipped into a certain liquor prepared
for the purpofe.

POMEGRANATE-TREE.

THIS tree grows both wild and cultured.
The branches of the firft are fmall, angular, and armed with thorns.   The bark
is red; the leaves fmall, like the myrtle;
and the flower is large, of a beautiful garnet,
and compofed of feveral leaves reprefenting
a little bafket of flowers.   The cup is oblong, purplifh, and in form like a bell.—
From this bloffom is produced a fruit, which
grows into a large round apple with a thick,
fmooth, brittle rind, adorned with a purple
cup.   This apple is called the pomegranate,
which is too well known in our elegant deferts

ferts to require a particular defcription.----
The wild pomegranate is only produced in
hot countries. The juice of the pomegranate is much valued in medicine. Of this
tree the Englifh reckon five forts, which are
cultivated more for ornament than utility.
They confift of the common, fweet, wild,
double flowered, and American dwarf pomegranate. The firft of thefe is the moft common in England, which, with care, has been
known to afford fruit that has ripened tolerably well in warm feafons: but as they
generally ripen late, they are feldom well
tafted. The double-flowered, continuing its
beautiful bloom for near three months, is
efteemed as the moft valuable flowering tree
yet difcovered.

RICE-PLANT.

THIS plant is much cultivated in the eaft,
and produces the grain fo much confumed,
which is called rice. Although a native of
the Eaft, great quantities of it have been
reared in South Carolina, where it is found
to fucceed as well as in its original foil : and
it being a grain that from its ufe may be called the manna of the poor, it has proved
moft beneficial to that province. The plant
bears its ftalk to the height of three or four
feet, and is much thicker and ftronger than
that

that of wheat or any other corn. The leaves are long like those of the reed, and the flowers blow in the top like barley: but the seed grows in clusters, and is enclosed in a yellow husk ending in a spiral thread. This plant growing in moist soils, where the ground can be overflowed with water, such as are desirous of cultivating it in Europe should place the plants, reared in a hot-bed, in pots filled with rich light earth, and placed in pans of water, which should be plunged in a hot-bed, and replenished as the water is by the heat diminished. In July they should be openly exposed, but in a warm situation, and with the same watery nourishment. Towards the latter end of August they will produce their grain tolerably ripened, if the autumn should happen to be favourable. Although rice be chiefly used for food, it is sometimes used in medicine. It nourishes well, stops fluxes, and is therefore found extremely serviceable in armies. As it increases blood, it restores in consumptions. The newest rice should be chosen, and such as is large, white, and well cleansed.

CORK.

CORK-TREE.

OF this tree there are feveral fpecies.—
The chief are the broad-leaved, the ever-
green, and the narrow-leaved with fmooth
edges. The firft is only requifite to be defcribed, which is always green, of a moderate height, refembling the oak, and having
a thick, light, fpongy bark, of an afh-colour, which is firft taken from the tree, and
afterwards feparated from the inner bark.
The leaves, cups, or acorns, refemble, like
the form of the tree itfelf, thofe of the oak.
It grows in Italy, Spain, and efpecially towards the Pyrenees and in Gafcony, &c.
The inhabitants of thefe countries, when
defirous of making a crop of this produce,
ftrip the bark from the top to the bottom of
the cork-trees, and pile them to a reafona-
ble height in a pit or ditch filled with water.
Having loaded thefe heaps with weights,
they leave them until they are thoroughly
foaked and ftraitened; then they are removed to another ditch, and from thence to a
third and a fourth. They are next taken
out of the water, dried, and packed in bales
for exportation. To choofe the beft cork,
the fineft boards that are free from knots and
chinks, of a moderate thicknefs, yellow on
both fides, and firm in texture, fhould be
felected. This beft fort of cork is called

the

the white cork of France, from its being chiefly produced about Bayonne in the province of Guienne. From the fame part is brought a fort which is called the Spanifh cork, which feems as if it had been burnt: but its blacknefs is faid to be caufed merely by having been fteeped in fea-water inftead of frefh water. The infide is, however, yellowifh, and eafily cut. Of this the thickeft fhould be chofen.

TOBACCO PLANT.

OF this production there are five fpecies; the firft is the Oroonoko, of which there are two forts; the one has very broad, rough, roundifh leaves; while the leaves of the other are narrow, fmooth, and pointed: but neither of them is valued by the planter, in confequence of their not being much confumed in England. The fecond fort is called the fweet-fcented tobacco, from its affording, when fmoked, a moft agreeable fcent; this fort is very much cultivated in Cuba, Brazil, Virginia, and feveral other parts of America; from whence it is fent to moft parts of Europe, but efpecially to England, where its general culture is prohibited, left the revenue fhould be diminifhed. The third fort is the greater narrow-leaved perennial tobacco, imported from the French fettlements in the
Weft-

Weft-Indies into the royal gardens at Paris, where it is cultivated in fmall quantities for the making of fnuff. The fourth and fifth forts are preferved in Botanic gardens, lefs for ufe than for variety.

Tobacco is raifed from feeds fown in a rich ground, where the rifing plants are covered, to defend them from the fun ; in the rainy feafons they are tranfplanted into large pieces of ground that are cleared and prepared for the purpofe. The diftance of the rows in thefe plantations is about two or three feet, or fuch a diftance as will not admit their extending leaves touching, which would caufe them to rot, by corrupting each other. The tobacco being thus tranfplanted, they only require to be weeded, until the flower-ftems appear, when they cut off the tops in order to afford more nourifhment to the leaves : the leaves hanging on the ground are likewife pulled fo as to let remain about ten or twelve upon each ftalk, which caufes a great increafe. The leaves, when ripened, are cut and fpread upon the ground : they are then ftrung upon certain cords in little knots, at fuch diftances as the plants may not touch one another : they are next hung to dry in the air, in a fituation guarded from the wet, during fifteen or twenty days. When fufficiently prepared, they are made into fuch forms as the purchafer defires.

COTTON

COTTON PLANT.

THE fruit of this plant is the cotton which is fo much ufed as a material of manufactures chiefly made at Manchefter. Its plant bears a ftalk about eight feet high, covered with a reddifh hairy bark, divided into feveral fhort branches. The leaves are rather lefs than thofe of the fycamore ; they are fhaped like thofe of the vine, and are fufpended by fmall ftalks adorned with a nap or hairy fubftance. The flowers are fine, large, and numerous, of a yellow colour mixed with red or purple, and fhaped like a bell : the flower is fucceeded by a fruit as large as a filbert, which, being ripe, opens into three or four partitions, where the cotton is found as white as fnow. Heat fwells each flake to the fize of an apple. There is another fort of cotton-tree that differs from the former in fize ; for this grows to four or five feet high : the flowers and fruit are like the former. Both thefe forts grow in Egypt, Syria, Cyprus, Candia, and the Indies. In Jamaica, Barbadoes, and other parts of the Weft-Indies, the cotton plant grows to a tolerable height, and fpreads on every fide its branches : it has fmall, green, pointed leaves, and bears a yellow flower refembling in form the rofe of the fweet-briar. The fruit is as large as a tennis ball, and has a
thin

thin crufty fhell, of a brown or blackifh colour. In thefe are found the cotton. In
fome of the American plantations there are
cotton bufhes very like thofe of Egypt, Arabia, &c.

MANDRAKE PLANT.

THIS plant is of two fpecies; one is the
common, and has a round fruit called
male mandrake; the other has a purple flower, and is called the female mandrake. The
leaves of the former rife immediately from
the root, and are about a foot long, and
broader than a man's hand, of a fmooth furface, a deep green colour, and of a difagreeable fmell. The flowers of both are
fhaped like a bell, which leave a foft globu-
lar fruit containing many feeds, fhaped like a
kidney. The root, according to fome naturalifts, reprefents the lower parts of a
man, and is therefore called anthropomor-
pha, which in Greek, fignifies the figure
of a man. But this feigned refemblance of
the human form is only devifed by the cunning of quacks and impoftors, who deceive
the ignorant by forming the frefh roots of
briony and other plants into thefe refem-
blances. There is likewife another ridiculous
fable devifed refpecting this plant; which is,
that as it is certain death to thofe who root it

A a 2   from

from its parent mould, the ſtem is tied to
a dog's tail, and thus it is taken from the
earth in order to prevent the above diſaſter
happening to any of the human ſpecies.——
The report of the mandrake crying like a
child, when torn from its ſoil, is equally falſe
and ridiculous ; for many of this plant have
been removed without any other effects than
thoſe attendant on the removal of all deep-
rooted vegetables. But what deſerves ·cre-
dit relative to the mandrake is, that the roots
will remain ſound above fifty years, and retain all the vigour of the moſt youthful plants:
they ſhould never be removed after their
roots have arrived to any conſiderable ſize, leſt
the lower fibres ſhould be broken, and thus
the growth of the plant be diminiſhed, and
its ſtrength debilitated ; if thus injured, they
will not recover their former vigour in leſs
than two or three years. Both the male
and female mandrake grow in hot climates,
and are moſtly found in plains. They are
propagated in gardens by ſeeds, which ſhould
be ſown upon a bed of light earth ſoon after
they are gathered. In this ſituation they
ſhould remain until the latter end of Auguſt.
Having kept them during this time free from
weeds, they ſhould be tranſplanted into the
places for their future vegetative exiſtence.
The ſoil of theſe ſhould be light and deep,
in order to admit the roots penetrating ſo
low into the earth as they are by nature formed to fix themſelves. Thus tranſplanted,
they

they will produce great quantities of flowers
and fruits for a feries of years. The mandrake is mentioned in the thirtieth chapter of
Genefis, where Reuben is faid to have found
one in the field during the wheat harveft :
it being faid in the Canticles, " The man-
" drakes give a fmell, and at our gates are
" all manner of pleafant fruit," feems as if
the fruit of the mandrake was delightful in
fmell; for furely Solomon muft mean a grateful fmell, otherwife he would never have
chofen it as an embellifhment of a paftoral
fong. However, the mandrake known to
us at prefent has no fuch delightful quality as
to render it fo valuable as to caufe a woman
to exchange her hufband, as Rachel did, for
one of them.

BALM OF GILEAD.

FROM the trunk of this plant flows a white
liquid balfam, which bears the name of the
vegetable. The plant bears leaves like rue ;
and white, ftarry flowers, which produce, in
their middle, berries enclofing a fmall kernel.
When the balfam firft runs, it is of the confiftence of oil of fweet almonds; but age
caufes it to refemble turpentine ; when it
lofes great part of its perfume, and turns
rather blackifh. When frefh, the fmell is
moft agreeably aromatic, and the tafte like
citron-peel.

citron-peel. Jericho was the only place where this balfam was to be found : but, since the Turks have posseffed the Holy Land, these shrubs have been transplanted into the gardens of Grand Cairo; where they are guarded, during the flowing of the balfam, by the Janissaries. At this time it is very difficult for the christians to obtain a sight of these balsams. With respect to the balsam itself, it is almost impossible to obtain any, unless from an ambassador, who may have some sent him, as a present, from the grand seignior, or from the soldiers appointed to guard this valuable liquid. This circumstance plainly evinces, that the balsam sold here, can only be the white balsam of Peru; which is prepared with spirit of wine rectified, or with some distilled oils. Mr. Pomet says he received from a friend, the present of an ounce, which he brought from Grand Cairo. He describes it to have been of a solid consistence, like the turpentine of Chio, of a golden colour, and a citron smell.

CEDAR OF LIBANUS.

THIS tree is very large, thick, and strait: the leaves are slender, and much narrower than those of the pine-tree: they are disposed in clusters along the branches; upon the upper

upper part of them grows erect the fruit, like our pine-apples; but they never drop in a whole ftate. It is faid there iffues from the trunk, in the warm months, a fort of white refin, which is very clear, of a grateful odour, and is called cedar gum: the large trees are faid to afford no lefs than fix ounces per day of this fubftance. The cones of the cedar, if preferved in time, will contain their feed for feveral years. They ripen moft commonly in the fpring, and are nearly twelve months old before they arrive to us from the Levant. To manage the cedar plant, we refer our readers to Miller's directions, in his gardener's dictionary.

What is mentioned in Scripture, refpecting the lofty cedar, cannot be applied to this tree; which, inftead of rifing in height, is more inclined to extend its branches in breadth. Mr. Maundrel obferves, that when he vifited mount Libanus, he only found fixteen large cedars remaining; but that there were feveral young trees of a fmaller fize. One of the largeft he found to be twelve yards fix inches in circumference, and thirty-feven yards in the fpread of the boughs. At about five or fix yards from the ground, it was divided into five limbs, each being as large as a great tree.

Cedar is faid to be proof againft the putrefaction of all worms, or animal bodies. The faw-duft is thought to be ufed by thofe

moun-

mountebanks who pretend to have the secret
of embalming. The wood is said, likewise,
to yield an oil which preserves books and
writings.

My Lord Bacon asserts, that cedar will
continue sound a thousand years. Of this
wood it is needless to observe, that the timber
work of that glorious structure, the temple
of Jerusalem, was formed.

ANANA PLANT.

FROM this plant is produced a species of
pine-apple that is reckoned, from its richness
of flavour, the king of fruits. It has the
delicious tastes of the peach, quince, and
muscadine grape, united. The top of it is
adorned with a little crown, and a bunch of
red leaves like fire. When the crown falls,
which is thought to be an emblem of its royal excellence, another succeeds, possessing
all its predecessor's qualities. The plant is her-
baceous, and has leaves somewhat resembling
those of the aloe. The fruit, which is like the
cones of the pine-tree, is supposed to have
been the cause of its name. The place of
its nativity is not determined: it was, however, first brought from the East-India fac-
tories, and planted in the hottest islands in
the West-Indies, where it succeeded so well,
as to afford now a most plentiful produce.

It

It has lately been introduced, with fuccefs, into the European gardens. The firft perfon who fucceeded in this attempt, was Monf. Le Cour, at Leyden, in Holland. From him, the gardens in England were firft fup- plied with this royal fruit. 'From its juice, is made a wine, almoft equal to' Malmfey fack; it will, likewife, intoxicate as foon, as the ftrongeft juice the grape affords.

GREAT AMERICAN ALOE.

THE aloe is a plant, which has leaves thick, and armed on the edges with fpines. 'The flower confifts of one leaf, which has fix parts at the top, like the hyacinth: the fruit is oblong, and divided into three cells; in which are inclofed flat and femicir- cular feeds. In the curious gardens of Botany in England, there are near forty different forts, which are natives of both the Eaft and Weft Indies : but the moft curious aloe is brought from the Cape of Good Hope. Moft of the African aloes produce flowers annually, when grown to a fufficient fize, which is often in the fecond, and feldom more than the third or fourth year after planting from off fetts: but the American aloes, which produce their flower-ftems moftly from the centre of the plant, feldom flower until they are of a confiderable age, and then but once

during

during the life of the plant; for the flower-
ftem, fhooting to fo high a ftature, draws
from the centre fuch a quantity of nourifhment as to render the leaves irrecoverably
decayed; and when the flowers are full
blown, fcarcely any of the leaves remain
alive: but whenever this happens, the old
root fhoots an innumerable quantity of off-
fetts, by which thefe plants are not only
preferved, but confidérably increafed.

The accounts of this plant are, like thofe
of many others, rather fabulous. That of
its blooming only once in a hundred years,
and making a report like a gun, are equally
falfe; for many American aloes have been
known to bloom in much lefs time. In the
year 1729, a great American aloe flowered
at the age of forty years, in a garden be-
longing to Mr. Cowal, at Hoxton: and of
a later date, fome have been known to
bloom at the diftance of twenty years.

SENSITIVE PLANT.

THIS plant is very furprifing in its con-
texture, and has caufed much invefti-
gation among the naturalifts, to account for
the contraction of its leaves when any of them
are touched. They clofe themfelves by
pairs, joining their upper fuperfices together. Aqua-fortis being dropped on the
fprig

fprig between the leaves was found to caufe them to clofe by pairs fucceffively to the top of each fprig, and to continue in this ftate fome time : but the next day the leaves on two or three fprigs were again expanded, except thofe on that where the aqua-fortis had been dropped, being withered from the place upwards, although they continued green downwards. A pair being fuddenly cut off with fciffars, the next pair above and below immediately clofed, and after a little time all on the fame fprig followed the example, which extended even to thofe on other fprigs. One of the harder branches being cut, emitted a liquor, which was very clear, and of a bright greenifh colour, bitter in tafte, and fomewhat refembling that of liquorice. The above experiments were made by Dr. Hook on fome fenfitive plants growing in a garden in St. James's park.

In the paffage of the ifthmus from Nombre de Dios to Panama, in America, there is related to be a whole wood full of fenfitive plants, which being touched, clofe their leaves with a rattling noife, and thus twift themfelves into a winding figure.

THE

SCIENCE of BOTANY

BRIEFLY EXPLAINED.

TO ufher our young readers into this
pleafing and inftructive fcience, we of-
fer the following compendium of botanical
illuftrations, to their attention, before they
proceed to the ftudy of the flowers we have,
in the following pages, fhortly defcribed.

Every fcience, except botany, poffeffes a
language peculiar to itfelf. Every perfon
who has pretended to teach, or explain, the
nature of plants, has chofen terms to exprefs
himfelf, according to his own caprice, or
his particular ftile of obfervation. This ar-
bitrary mode of treating botany, has confiderably bewildered the ftudent; and even,
fometimes, diffuaded him from purfuing the
fcience with that avidity and pleafure he
would otherwife have done. Although the
vocabulary of botany has been always fub-

jeĉt

ject to this variation, it has never experienced more innovation than of late years: but, notwithstanding we lament this deficiency of stability in botanical language, we are happy to find that, sometimes, the alterations have been very judicious amendments of terms falsely used by the ancients; for the modern botanists have named the plants from the parts which they contain; while their predecessors have named them from outward appearance, or supposed qualities. Thus are the long terms, and denominations, which only perplexed the mind, and burdened the memory, abandoned. Conformably to this improvement, Linnæus proposes simple and proper terms, to express not only the different parts of plants, but, likewise, their forms, qualities, situations, directions, and mode of existence of each part respectively. This method has, in general, been adopted by all succeeding writers in this science.

No method could be so proper for classing plants, as that adopted by Linnæus; namely from their sexual difference. This is most natural, and least subject to variation, from the difference being described according to the variation of the stamina in the male, and the pointals in the female parts of a plant.

According to modern botanists, plants are described as consisting of six parts:—the root, *radix;* the trunk, *truncus;* the support,

port, *fulcra* ; the leaves, *folia* ; the flowers,
flores ; and the fruit, *fruĉtus.*

———————————

1. RADIX—*The* ROOT,

IS that part of the plant which adheres to
the ground, from whence it draws its nourifhment.

Roots are eithet fibrous, bulbous, or tuberous.

The fibrous root is either perpendicular,
horizontal, flefhy as the *carrot*, hairy as the
roots of *grafs*, or branching.

Bulbous roots, (among which are the
fnow-drop, hyacinth, and tulip) are either
folid, as the *turnip*; coated, as the *onion* ;
fcaled, as the *lily* ; double as the *orchis* ;
or cluftered, as the *white faxifrage.*

Tuberous roots are compofed of many
flefhy tubers, as the *garden ranunculus* ;
and either adhere clofely to the ftalk, or are
fufpended from it by threads.

———————————

2. TRUNCUS—*The* TRUNK,

RISES immediately from the root, and
fuftains the branches. This part is called
a trunk in trees, and a ftalk in plants.

Stalks

Stalks are either fimple, or compound.

A fimple ftalk grows from the root to the top, as the fun-flower; and is diftinguifh-ed by its either being naked, leafy, upright (as the lark's-fpur), oblique, twining, pli-ant, reclining, lying on the ground (as the nafturtium), creeping (as the Panfy), having roots as long as itfelf; living feveral years, or. only one year; being woody, fhrubby, cylindrical in form (as the ftar-flower); having two, three, or more an-gles; and being ftreaked, furrowed, or channeled, fmooth, rough (as the after), hairy, or prickly (as the rofe).

A branching ftalk is one that fhoots late-ral branches, as it afcends, as the wall-flow-ers; and is diftinguifhed by the branches being either irregular, large, numerous (as the piony), fupported, prolific in leaves, fruit, or flowers (as the lily of the valey, and the jonquil).

A compound ftalk is one foon divided into branches, as the flower of Parnaffus; and is diftinguifhed by being either forked, having two ranges of branches, or having thefe ranges fubdivided; tubular like a ftraw; being entire, branched, uniform, jointed (as a pink), fcaly, or with or with-out leaves.

3. FUL-

3. FULCRA—*The* SUPPORT,

IS that part which fuftains or defends certain parts of a plant, and is divided into the
following ten kinds; the leaf fupporting the
flowers, the tendril or clafper (as the honeyfuckle and fweet-pea,) the fpine, the thorn,.
the footftalk of the leaf, the footftalk of the
flower or fruit (as the columbine,) the general ftalk, the gland, and the fcale. Each
of thefe have their fubdivifions, which we
omit, as being too minute for the attention
of young ftudents.

---

4. FOLIA—LEAVES,

ARE divided into three claffes, of
fingle, compound, and determinate.
Single leaves are thofe that have footftalks
fupporting only one, as the cyclamen ; and
are defcribed according to their circumference ; border, furface, fummit, and fubftance.

Their circumference and border are either
round, nearly round, oval, reverfed oval,
oblong, fhaped like a wedge, angular, fpear-
fhaped (as the belvidere,) narrow, fhaped
like an awl, triangular, deltoide, or having
four corners, quinqueangular or five-cor-
nered,

nered, fhaped like a kidney, a heart, a
moon, an arrow, or a pike, divided into two
or three parts, formed like a hand, pointed
like a wing, jagged, indented (as the tuberofe,) divided or not into parts, fingly or
double fawed, notched, grifly, ciliated or
hairy like an eye-lid, lacerated, or feemingly
torn or bitten, curled, or entire.

Their furface is diftinguifhed by being
either downy, foft as velvet; hairy, as the
fox-glove; ftinging; rough; fmooth, as
the daify; briftly, prickly, warted, polifhed, plaited, waved, wrinkled; veined, as
the gilliflower or carnation; nervofe; plain,
as the auricula flower; depreffed, compreffed, convex, concave, or channelled.

Their fummit, or top, is either truncated,
blunt, as if bitten, hollow, obtufe, pointed
(as the amaranthus,) fhaped like an awl, or
taper like a pillar.

Their fubftance is either hollow, flefhy,
or membranous (as pinks.)

Compound leaves are either fimple or decompound.

A compound leaf is formed of feveral
fmall leaves growing from one footftalk, and
is confidered as one whole, produced from
a fingle compofition, as the ranunculus,
rofe. carnation, pink, &c. They are either
fingered, compofed of two, three, or many
leaves, refembling wings expanding from
their common footftalk, and having alternate
leaves, or being doubly winged.

A decom-

A decompound leaf has a footſtalk dividing twice or more times before it is garniſhed with leaves.

Determinate leaves are diſtinguiſhed by their direction, place, inſertion, or ſituation.

The direction is the manner in which the leaf expands from the bottom to the top, and is either arched, upright, ſpreading, horizontal, reclining, or revolving backwards.

The place is determined by the part of the plant where it is faſtened, and is either called the ſeed leaf from riſing immediately from the ſeed, or radical from riſing firſt from the root.

The inſertion is the manner in which a leaf is faſtened to plant, and is either faſtened to the diſk, or has a footſtalk to its baſe, grows from the branch without a footſtalk, is faſtened by a membrane, or ſurrounds the ſtalk: without any part of the border adhering to it, like the hare's-ear.

The ſituation is confidered from the poſition of each in relation to the others. The ſituation is, therefore, either jointed, ſurrounding the ſtalks like ſtars, oppoſed to each other (as the jeſſamine,) growing in an alternate poſition on each ſide their footſtalk, or without any order, cluſtered (as the flowers of the ſweet William,) ranged like the tiles of a houſe, or the ſcales of a fiſh.

5. FLO-

5. FLORES.—*The* FLOWERS.

THE flowers of plants are divided into four parts : the cup, *calyx ;* the petal, or flower-leaf, *corolla ;* the ſtamen, *ſtamina ;* and the pointal, *piſtillum.*

The CUP OF THE FLOWER is that which incloſes, and ſuſtains the flower ; and is divided into ſeven ſorts ; the *perianthium, involucrum, ſpatha, gluma, amentum, calyptra,* and *volva.*

. The *perianthium* is the moſt common of the flower-cup ; conſiſts often of many parts ; ſometimes of only one part, ſeparated half-way into ſeveral diviſions, as the India pink ; and always ſurrounds the bottom of the flower.

The *involucrum* embraces many flowers collected together, and which have each of them a perianthium.

The *ſpatha* is a ſheath, which covers one or more flowers, that are generally without a perianthium ; it conſiſts of a membrane, faſtened to the ſtock ; and differs in its figure and ſubſtance.

Gluma is a ſort of chaff, which particularly covers grain and graſs ſeeds.

The *iulus,* or *amentum,* is a maſs of male or female flowers covered with ſmall ſcales, and faſtened to an axis, in the form of a rope, as the irregular flowers of the violet.

The

The *calyptra*, or *coif*, is a thin, conical, membranous cover to the parts which generate fruitage.

The *volva*, or *purse*, is a thick covering inclofing feveral fpecies of mufhroom productions.

The COROLLA, petal or flower-leaf, is one of thofe which form the flower, and furround the generative parts of the plant itfelf. Of thefe there are the *petal*, and the *nectarium :* they are either entirely one, as the convolvulus, or formed of many pieces. The petal is generally diftinguifhed by the beauty of its colour, and the nectarium by containing thofe fweet juices which the bees change into honey. The corolla is fometimes without a footftalk, as the martegon.

The STAMEN is the male part of flowers, and confifts of the *filament* and the fummit or *anthera*, as the paffion-flower.

The *filament* fuftains the anthera, apex, or fummit, and is either formed like a thread, or fhaped like an awl.

The *anthera, apex,* or *fummit,* is the effential part of the ftamina, and contains the male organ of generation. It confifts of a little bag, of one or more cavities, containing the male farina.

The POINTAL includes the female parts of flowers, and confifts of the *germ, ftyle,* and *ftigma.*

The

The *germ* inclofes and defends the feeds.

The *ftyle* rifes from the germ, and fupports the ftigma.

The *ftigma* is the female organ of generation, and is fituated upon the top of the
ftyle, if any ; if not, it fits upon the germ.

---

6. FRUCTUS—*The* FRUIT.

THE different fpecies of fruit, fuch as
plums, berries, apples, feeds, &c. are too
well known to require a defcription.

The CLASSES.

FLOWERS are either hermaphrodite,
from having both the fexual diftinctions of
male and females, ftamina and pointals ;
male, from having *ftamina* only ; or female, from having only *pointals*.

The *ftamina* are either detached from
each other, united together by one of their
parts, or joined fometimes with pointals :
they are of equal length, or have fome
fhorter than the reft ; and the number, proportion, and fituation of the ftamina determine the *claffes*, as the differences of the
pointals determine the *orders* of flowers.

The

The claffes, according to the number of ftamina in the male parts of the flower, are called,

1. *Monandria*, one ftamen.
2. *Diandria*, two ftamina.
3. *Triandria*, three.
4. *Tetrandria*, four.
5. *Petandria*, five.
6. *Hexandria*, fix.
7. *Heptandria*, feven.
8. *Octandria*, eight.
9. *Enneandria*, nine.
10. *Decandria*, ten.
11. *Dodecandria*, eleven.
12. *Icofandria*, when more than twelve.
13. *Polyandria*, when more than thirteen.

Thofe flowers which have two ftamina fhorter than the reft, are called,

14. *Dynamina*, as having two long and two fhorter ftamina.

15. *Tetradynama*, as having four long and two fhorter ftamina.

Thofe flowers which have their ftamina united together or with a pointal, are thus diftinguifhed.

16. *Monadelphia*, ftamina united into one body.

17. *Diadelphia*, ftamina into two bodies.

18. *Polyadelphia*, ftamina into three or more bodies.

19. *Syngenefia*, the ftamina forming a cylindrical body.

20. *Gyn-*

20. *Gynandria*, the stamina fitting upon the pointals.

Those plants of different figures are thus distinguished.

21. *Monoecia*: the plants of this class have male and female flowers upon the same individual.

22. *Dioecia*, have male and female flowers on different individuals.

23. *Polygamia*, have hermaphrodite flowers upon the same individual.

ORDERS.

THE orders, or subdivisions, of the classes, are distinguished by the pointals, or female parts of the plant or flower, as the classes are by the stamina, or male parts of the flower. The number of pointals or stigmas are counted.

The chief distinctions are the number of pointals, and nature of seeds, the nature of pods, and the number and gender of the florets. According to the number of the pointals, the orders are termed monogynia, digynia, &c. according to the nature of the seeds, gymnospermia, angiospermia; according to the pods, siliculosa, siliquosa; and according to the number and gender of the florets, they are termed polygamia æqualis, polygamia superflua, &c.

A CON-

A

CONCISE HISTORY

OF

FLOWERS.

JONQUIL.

THIS charming flower comes, with all its
graces, to deck the spring ; it confifts of feveral fpecies ; but the great jonquil has a ftem,
about a foot in height, which bears from a
third part upwards, feveral golden bloffoms,
confifting of five or fix leaves, all curling in
a moft agreeable and beautiful manner. It
is multiplied by feed; but, more properly, by
their bulbs. They require a good, but not
a very rich foil; and are ufually planted along
the borders; thus affording a moft agreeable
embellifhment to the walks and parterres of
any garden meant to be diftinguifhed for its
tafte and elegance. .

ANEMONE.

ANEMONE.

THIS beautiful flower, with proper culture, will blow twice a year; and thus continue to grace our gardens, when they are abandoned by all the reft of the flowering tribe. Their colours are chiefly red, blue, and purple. The root of thefe plants fhould be taken out of the ground, and preferved like thofe of the ranunculus. They grow beft in a fandy foil.

When the feeds crack, or fhew their down, they fhould be gathered, to prevent their being difperfed by the wind. From thefe feeds, innumerable varieties may be raifed: and if they are fown in February, and lightly covered with earth, they will blow the fecond year after fowing.

LILY.

THIS flower is a great ornament to a garden. The noble height of its ftem, and the fimple grandeur of the flower, render it a moft delightful fpectacle to thofe who have the leaft tafte for the beauteous productions of nature. The lily is too well known, and admired, to require any particular defcription of its form or colour. The culture requires

no

no curious rules, from its being eafily reared in any foil : and, as if nature meant this charming flower fhould be enjoyed by the poor as well as the rich, we find it thrive with the leaft attention. Such is the beauty of the lily, that many European noblemen place them in pots, in order to decorate the avenues to their fumptuous palaces.

Some garden-walks are entirely bordered with them : and, indeed, wherever they are placed, they are always beautiful.

LARKSPUR.

THE larkfpur is one of thofe flowers that feem to delight in difplaying the variety of colours with which the flowers of each ftem are decorated. They grow on ftalks, of three feet high ; and, when choicely reared, afford, in a bed, one of the moft beautiful fpeftacles that Flora has to prefent, for our delight, wonder, and contemplation. It is generally fown in February ; and may be expefted to bloffom, in all its richnefs of fplendid beauty and elegance, in June and July. If properly attended, they will continue their bloom until Auguft, or September.

DAFFODIL,

DAFFODIL, or LONG-NECKED NARCISSUS.

WHICH is called *cou de chameau*, i. e. camel's neck, from the long ftalk, when charged with flowers, reprefenting the neck of this animal. This flower is to be admired for its being an agreeable ornament to the rural parts of a garden. They bloffom in the fpring, and grow about a foot high. The daffodil thrives beft in a rich foil, with which the bulbs need only be covered; it fhould not be much expofed to the fun, from the flower deriving moft beauty from the latenefs of its appearance. The bulbs fhould be fet about four fingers diftant from each other, in order to afford fufficient room for their expanfion. It fhould be removed every three years. They flower in March.

COLCHICUM, or MEADOW SAFFRON,

IS fo called from its growing in Colchis, a country in the neighbourhood of the kingdom of Pontus, famous for the fable of the golden apples, and the golden fleece. It is faid to be fo ftrong a poifon as to kill dogs, from which quality it is called dog's bane. Of the meadow faffron there is a variety of fpecies.

fpecies. Its general defcription is, being a plant that fhoots from its root five or fix oblong leaves, about an inch broad, fmooth, and of a brownifh green. Amid thefe leaves rifes the ftalk, bearing at the top a yellow fingle-leaved flower like a pipe, and cut into fix parts. The Colchicum will grow in any foil. It is multiplied by bulbs, which are produced every year in abundance. They fhould be planted in pots or borders, and tranfplanted in July; in which ftate they fhould lie until September. They flower in March.

POLYANTHUS

IS divided into the primrofe and cowflip kind ; and thefe are fubdivided again into the fingle-flowering, double-flowering hofe in hofe, pentaloons, and feathers. The fingle-flowering are chiefly white, yellow, red, purple, and violet-coloured. They are multiplied by feeds, fown in February, upon a place prepared with earth taken out of decayed willows ; often refrefhing the new-fown fpot with water ; and keeping it fhaded from the fun, all April and May, until the young plants appear. The Primrofe kinds bloffom clofe to the ground ; and the Cowflip fpecies, about fix inches higher. Both thefe-forts may be planted near the edges of borders,

and

and near houfes, for the enjoyment of their
agreeable fmell.　Nothing can be more delightful than a number of thefe Flowers, ac-
companied with violets, growing under
hedges, in avenues, and artificial wildernef-
fes.　They flower in April.

PERSICARIA

HAS a towering ftem, about five feet and
a half high, refembling a Sugar-cane, which,
towards the bottom, is garnifhed with feveral large green leaves, like thofe of lilac.
It has a garnet bloffom which grows in the
form of a feather, that hangs from their
ftems with confiderable grace and beauty.
They are cultivated in moft gardens diftinguifhed for their choice affemblage of elegant
flowers.　Their time of bloffoming is during the fummer months, when the parterres
of thofe gardens in which they are cultivated, derive confiderable ornament from their
beautiful and fingular appearance.

TULIP.

THE tulip requires nothing but a fine fcent,
to render it the fineft flower in the world.
Their infinite varieties difplay fuch beauties

as

as eclipfe every other pride of the garden. Thefe ornaments of nature are as kind as they are beautiful ; for they continue regaling the fight with a fucceffion of their charms, from March to the latter end of May. They are divided into claffes ; the early and later blowers. Their varieties are chiefly diftinguifhed by the names of cities, or fuch like characters. A good tulip is known by its towering ftem, its beautiful colours ; with a flower fhaped like an egg, without fharp points to their petals ; but what renders them the moft valuable, is their variety.

The flower-ftems, being left upon the roots, will perfect their feeds about July. The feeds are gathered when they begin to crack.

JERUSALEM CROSS.

THIS flower is a fpecies of the Lychnis ; and it is called by botanifts, *Flos Conftantinopolitanus*, from being originally brought from Conftantinople. This plant fhoots into feveral ftems, about two feet high ; and divides itfelf into different branches. The leaves are long and pointed, of a green and brown colour. On the top of each ftem grow the flowers, confifting of five leaves, which hang down, like the tops of fennel,

and

and reprefent little croffes, fometimes of a white, but more generally of a fcarlet colour. They have an agreeable odour. The Jerufalem crofs will thrive in any fubftantial foil; but it grows beft in the fhade. The culture is the fame as of the Lychnis; to which we refer our readers. It flowers in July; and is reckoned a great ornament, among any others you may pleafe to plant it. Care fhould be taken to water it, in hot and dry feafons.

NARCISSUS.

OF this flower there are feveral fpecies; but as the narciffus polyanthus is one of the moft early bloffoms, we fhall briefly defcribe it. Its feent is fo fweet, that many confider it not lefs defirable than the Jonquil. This, like all the other narciffufes, fhould be propagated from offsets, taken from their roots.

The polyanthus is greatly admired for its fplendor and variety of colour, in both of which it has no fmall refemblance to the auricula. In the rural parts of our gardens, thefe, as well as the daffodil narciffus, are a very agreeable ornament; which has caufed them to be frequently mentioned by the moft eminent of paftoral writers.

FRITIL-

FRITILLARY

IS a plant that has a ſtem about a foot high, round, ſmooth, and of a deep green colour. It is garniſhed with about ſix or ſeven leaves, placed irregularly, and which are long and narrow. At the top of the ſtem grow one or two flowers, hanging down in the ſhape of a bell : theſe are ſpeckled with ſeveral colours, and are compoſed of ſix leaves. The colours, being placed in the form of a chefs-board, have cauſed this plant to be called the Fritillary, from *Fretillus*, which ſignifies a chefs-board. Fritillaries are multiplied by bulbs and ſeeds. The bulbs are planted in September. They ſhould be placed three inches deep, and at the ſame diſtance from each other. They flower in April.

JESSAMINE.

ALTHOUGH all the ſpecies of Jeſſamines grow in a very irregular form, and are never ſubmitted to the pruning-knife, they are a beautiful ornament to any garden. Of the Jeſſamines, there are too many ſorts to be here deſcribed ; we ſhall therefore confine ourſelves to the common jeſſamine, which is ſo great a decoration to our gardens.

It

It is a fhrub that fhoots forth feveral fmall branches; which are adorned with leaves oblong, pointed, placed in pairs along each branch, which terminates with a fingle leaf: at the end of the branches grow the bloffoms, in form of umbrellas, confifting of five delicate white leaves, which poffefs a moft agreeable fmell. When the Jeffamine is in bloom, nothing can be more pleafing than the contraft of the green ground with the ftarry flowers with which it is fo numeroufly ftudded.

CARNATION.

THESE are called, by the Greeks and Romans, the white violet, from being of the fame fpecies with refpect to the flowers. The Gillyflower is reckoned one of the moft principal ornaments of our gardens. The variety and great number of its flowers feem to have acquired it this diftinction. The leaves of the ftem refemble thofe of fage: from the middle of the root, the ftem rifes about eighteen inches, and then runs into feveral branches, tufted with beautiful flowers, compofed of four leaves, in the form of a crofs, which have a moft fragrant fmell. This plant is raifed from feed fown in March, in hot-beds, in fmall drills drawn acrofs each other: the feed being fown, is

covered,

covered, with the hands, as lightly as poffible. When the plants appear, they muft
be fecured from the froft by glaffes, matting, or dry dung. Among the gillyflowers
is ranked what is commonly called the carnation, old blowers, &c.

PASSION FLOWER.

THIS flower cannot be efteemed lefs than
a miracle, fince God has thought proper to
defcribe on it the principal emblems of the
death and paffion of our Saviour. The
leaves are pointed, like a crown of thorns:
the whitenefs of the leaves reprefents the in-
nocence of Chrift; the red ftrings are emblems of his being fcourged; and the little
column, in the middle of the flower, is
thought by divines to be the figure of the
pillar to which our Saviour was bound:
another part reprefents the fponge; and the
ftamina, growing over the pillar, remind us
of the three nails with which he was nailed
to the crofs, and, in a word, the pointed
leaves raife a perfect idea of the fpear with
which his facred fide was pierced. This
moft curious flower grows in all forts of
ground, efpecially in a foil inclinable to
moift rather than light; it is multiplied by
roots fet three inches deep. As the roots
fpread confiderably, care fhould be taken to

D d

prevent

prevent their injuring the roots of other neighbouring flowers.

AMARANTHUS

IS a plant that has, rising from its root, leaves that are large, pointed, of a brownish green, bordered with red. From the centre of these leaves grows a stem about eighteen inches high, of a red colour, bearing flowers either of a violet, purple, crimson, orange, red, or scarlet colour. From the beauty and simplicity of these colours, the amaranthus is always esteemed as a most valuable appendage to a garden. The seed, which is remarkably small, curious, and beautiful, is preserved in little boxes until the winter. These flowers appear graceful in pots filled with kitchen-garden earth and bed mould. If watered constantly and carefully, they will grow, in this state, to a fine size, and will make a most beautiful appearance: and, as the flowers continue a considerable time, and flourish when other flowers are scarce, the amaranthus is considered as no inconsiderable part of an elegant garden.

ROSE.

R O S E.

ALTHOUGH rofes are generaly ranked among flowering fhrubs, yet, as they are reckoned among the greateft ornaments of a garden, and are the chief beauty of any affemblage of flowers, we fhould think ourfelves remifs, in omitting a brief account of them, in this fhort defcription of flowers.

As a general defcription of the many forts of rofes,—they grow on fhrubs, that fhoot forth hard, woody, thorny branches; with oblong leaves, indented, and armed with prickles. On thefe branches grow the flowers, confifting of leaves, in a round form; their cups are leafy, and turn to round, or oblong pulpy berries. The pale rofe is fair, large, of a carnation colour, and poffeffes an agreeable fmell and appearance. The damafk rofe is a fmall, white, fingle or doule rofe, with a mufky fcent. The common white rofe is large and beautiful; and remarkable for being, with the red rofe, worn as the diftinction of the houfes of York and Lancafter. The yellow rofe has broad leaves, of a lemon colour, without fmell. The monthly rofe is like the damafk, and has red flowers, growing in bunches. The ftriped rofe has white and red ftreaked leaves: and the mofs rofe is fo called from the ftem and outward leaves appearing to be
covered

covered with mofs, in a manner that appears fingularly beautiful.

RANUNCULUS.

THE ranunculus, next to the tulip, is defirable for its beauty. There are feveral forts of them imported into England every year from Turky. This plant blooms in April and May upon ftalks about fix or eight inches high. The double flowering forts are crowded with petals, like Province rofe flower. The colours of them are deep fcarlet, veined with green and golden hues, yellow tipped with red, white fpotted with red, orange colours, plain white, yellow with black, and one fort of a peach-bloom colour. The fingle ranunculus blows fomewhat taller than the double, and is moft agreeably variegated with pleafant colours. They are both increafed by offsets, found about the roots, after taken from the ground. They may likewife be propagated from feed, faved from the fingle bloffoms. The Englifh are indebted chiefly to the French for them, in confequence of their climate being too cold for their culture.

DAISY.

D A I S Y.

THE daify, being of an agreeable afpeƈt, was called by the Romans, *bellis*, from *bellus*, *i. e.* handfome. The daify has fmall, oblong fmooth leaves, both intended, and otherwife : in the middle of thefe leaves rife little, long ftalks, tufted with a radiated flower, which is fometimes white, red, and variegated.

The daify, for its fimplicity of beauty, and being the early grace of our banks and mea-dows, has been ever, and juftly, one of the moft charming fubjeƈts of paftoral poetry. To gather them, is the firft pleafure of lifping infancy; and to view them, is the firft delight of the humble cottager. Although this plant produces feed, yet thofe who cultivate them in their gardens, replant the fplit roots. It, grows very low ; and is a moft proper and beautiful border, either in the flower or kitchen garden.

T U B E R O S·E

IS a fort of hyacinth, called hyacinthus indi-cus. Although this plant is from fuch a dift-ance as Afia, yet it is now plentiful in moft parts of Europe. The tuberofe has, growing from its roots, feveral leaves, about fix inches

P 2

long, ftrait, and pointed at the end. In the
middle grow a ftem, to the height of three
or four feet, and about half an inch in diameter.   On the top of the ftem grow the
flowers, like lilies, fingle-leafed, fhaped like
a pipe, indented, and looking like a bell.
The flowers blow fucceffively, which caufes
the tuberofe to continue long in bloffom.
So fweet is their odour, that they perfume
the place wherein they are fet. This plant,
if fet in May, will flower in Autumn. They
fhould be placed where the fun is hotteft.
They will be found a greater ornament to
windows than to parterres.

SNOWDROP.

ONE of the firft offerings whichFlora difplays on the fhrine of nature, is the fnowdrop.
Pallid, like the cheek of fpring, are its leaves;
and, like the feafon in which it appears, its
bloffom hangs languid on the verdant ftem.
The flower is compofed of fix leaves which
together form a bloffom, fimilar in fhape to a
bell : the odour is as grateful as the colour
is delicate. The fnowdrop, being a bulbous
plant, is raifed from its root, and is generally
ranged with the narciffus. Although it is a
common flower, yet fuch is its beauty, fimplicity, and cheering appearance, that it
generally accompanies the crocus in all parterres

terres diftinguifhed for their variety or their
elegance.

SWEET-WILLIAM.

THERE are two forts of this plant,
confifting of fingle and double flowers. The
fingle fort only differs in the colour of the
flower : the one has branches of bloffoms
variegated with red and white : the other
has clufters of deep crimfon-coloured flowers. They both bloffom in June and July, upon ftalks two feet high. The double fort
produces its beautiful red flowers in the fame
months, but upon fhorter ftems. The fingle-
flowered fweet William may be raifed from
feeds fown in March : They will bloffom the
fecond year. The double fort is propagated
from flips, taken from the root in March or
April : if planted in a loamy foil, they will
thrive the beft. The others may be alfo
increafed by the fame means, or if they are
laid down in the earth like carnation layers.

CYCLAMEN.

THE cyclamen is fo called in Latin, French
and Englifh, from the root being almoft
round. It is a plant that produces from the
root,

root, leaves that are broad, almoſt round,
of a dark green colour, ſpeckled on the
outſide, and with purple on the inſide : In
the middle grow long pedicles, and at the top
of which are the ſingle-leaved flowers,
dividing into five parts, folding inwards.
Autumnal cyclamens bear a red flower,
ſweetly ſcented. In this ſeaſon, blows one
called the Conſtantinople cyclamen, which
bears the firſt year twenty flowers ; he
ſecond fifty, and the third two hundred,
and all without the leaſt ſmell. The cyclamen is raiſed by ſeeds. The autumn cyclamen ſhould be ſown in autumn, and the
ſpring cyclamen in the ſpring.

SCARLET LYCHNIS.

THE beauty of this plant is ſuch, as to
cauſe it to be ranked among the moſt elegant
parterres. Both the ſingle and double lychnis are very delightful in appearance, they
bear bunches of ſcarlet flowers, upon ſtalks
above two feet high, in June and July. They
are ſo greatly eſteemed, that gardiners rear
them in pots, to decorate the moſt beautiful
parts of their garden, or to be placed, in the
ſummer ſeaſon, in chimnies, where they prove
a moſt pleaſant ornament. The double kind
is increaſed by ſlips, taken from the root in
March. The ſingle flowering kind may be

propagated

propagated by the fame means, or raifed in March from feeds, which bloffom the firft year. An open fituation, and a light foil, are moft proper for their cultivation.

CROCUS.

THIS early flower, as if anxious to fhare with the fnowdrop in cheering the departing gloom of winter, appears in January and February, but not to be a mere fpectacle of beauty : it produces a moft ufeful fubftance, which is faffron. The fhape of the flower refembles the lily. It poffeffes an agreeable fcent. Confidering its cheerful afpect, when few flowers appear, and its producing fo valuable an effence, it is rather a wonder it fhould not be more cultivated in our gardens. The true crocus is rather to be multiplied by the root than by its feed. It requires a rich foil, and ought to be planted in a ground expofed to the foftering rays of the fun.

COLUMBINE.

THIS plant is called aquilegia, from *aquila* an eagle, in confequence of the leaves of its flower being hooked like the beak and talons of that bird. The columbine fhoots indented

leaves

leaves of a blueifh green, and growing to long
ftalks. In the middle, rifes a ftem of eighteen
inches long, which is flender, and of a reddifh
colour : from this ftem fprout feveral little
fprigs, which fupport a flower compofed of
five flat and five hollow leaves, coloured
with red, blue, white, chefnut, and carnation. Columbines require a rich foil, and are
cultivated by fowing the feed very thinly in
September, in beds well dug, where it remains until the plants are ready to be removed to the plots of a parterre. The columbine is one of thofe lafting plants which
is kept alive by its roots, and will live a long
time in the earth without requiring to be
fown again.

DOUBLE MARYGOLD.

THIS plant has been admitted into our gardens, from the richnefs of the colour, and
the beautiful form of the numerous leaves.
Nothing can be more fplendid than their golden hue. With refpect to the difpofition of
the leaves, they feem as if Flora had particularly difpofed them into the form of a crown,
for her own embellifhment. The leaves are
not only beautiful in themfelves, but they are
allowed, by phyficians and botanifts, to poffefs great medicinal virtues : they are faid
to cheer the fpirits, by their infufion, as
                                    much

much as they cheer the fight by their appearance. Their flavour is likewife fo agreeable, as to have caufed it to have been mixed among the herbs that are ufually boiled in our broths and foups. Thus after delighting us in the parterre, they heighten the delicacies of our table.

BELVIDERE.

FROM the leaves of this plant, refembling thofe of flax, it is called in Latin, *linaria*, from *linus*, which fignifies flax. It rifes into feveral ftems, two, three, or four feet high; and fhoots into many branches, garnifhed with ftrait, oblong leaves, of a light-green colour. At the extremities of thefe boughs appear fingle flowers with irregular leaves. Thefe plants are of ufe in little courts, where they are fet two feet diftant from each other, in borders raifed for the purpofe; or in pots, placed in fymmetrical order. The belvidere is multiplied by feed, fown in plain ground, in any part of a nurfery; from whence it is removed, as foon as it is ftrong enough to be replanted. As the air injures the root, it fhould be replanted the moment it is taken from its native foil, and watered immediately.

PRIM-

PRIMROSE.

THIS flower very early graces the lap of nature. Its golden leaves are frequently feen rifing from the fnowy beds. So welcome is this flower to man, that in Europe it is frequently reared in pots; which are placed to adorn the windows, when fcarcely any verdure is to be feen abroad. When planted, it fhould be placed in good garden mould, and in a warm fituation, among the fmalleft flowers, or elfe to edge the compartments of the parterres with its golden tiffues. As no flower is more cheering, or agreeable to the fight, it generally graces the moft choice and beautiful gardens.

FLOWER OF PARNASSUS.

THIS plant is called parnaffia, or gramen parnaffi, by the botanifts, from its being found on the mountain of Parnaffus. It bears leaves very like thofe of the violet; from amidft thefe leaves rife feveral ftems, about fix inches high: on the top is a rofy flower, compofed of feveral unequal leaves, fringed, and difpofed in a circle. This plant is annual, and confequently multiplied by feed, which fhould not be thrown too thick. It thrives beft

in

in a fat, moift earth ; and is cultivated like
thofe other plants that are fown in hot-beds
in March, and which are confequently to be
fecured from the cold by glaffes, ftraw or
matting. This flower is not only a great
beauty in parterres, but in pots, or very
large tubs, where it appears to equal advantage.

WALLFLOWER

IS called by fome, the yellow gillyflower.
It confifts of both fingle and double flowering kinds. It fhoots out leaves of a dark green
colour, that are pointed at the end : between
thefe leaves, grow feveral branchy ftalks ; on
the top of which, appear the flowers, compofed of four, and fometimes more leaves,
of a yellow colour. The fingle wallflower is
multiplied by feed, and the double by layers,
or flips.

This flower will grow every where; even
upon walls, or among rubbifh : but, when
cultivated, more care fhould be taken of
them, as they will prove an agreeable orna-
nament to borders, or any other parts of a
garden not deftined for more choice flowers.

E e   BLUE

BLUE BELL.

THE blue bell plant shoots forth stalks two feet and a half high, which are hairy, and furnished with leaves: these are oblong, broad, and pointed at the end, notched at the edges, and downy; along these stalks, and at the stems of the leaves, the flowers grow, in form of bells: these blossoms are blue, notched at the brims, and divided into four parts; each is supported by a calyx, or little cup, divided likewise into five parts. This flower delights much in the soil of a kitchen garden. It is multiplied by sowing the feed, as thinly as possible, on the end of a plot well dug, and smoothed on the surface. The time of sowing is September and October, and that of flowering is July.

SUNFLOWER.

THIS plant is called turn-sol by the Italians, which turning towards the sun: it is therefore called turnsol by several of our botanists. The cause of its turning towards the sun, is from the flower being heavy, and consequently inclining the stem to that position it is liable to, from being warped by the rays of this luminary.

The

The funflowers are of two forts : one produces a ftem between five and fix feet high, which is very ftrait and branchlefs, with leaves nearly as large as thofe of the vine, jagged, pointed and rough : on the top of this ftem appear the flowers, refembling the fun. Care fhould be taken in what part of a garden it is planted, left it fhould choke the flowers growing near it. The places moft proper, are the broad allies planted with trees, and between which the turn-fol may be planted at three feet diftance.

INDIAN PINK.

ALTHOUGH this plant has a ftrong fmell, yet it is raifed in our gardens, for its beautiful flower. The Indian pink fhoots into a ftem, about eighteen inches high, and then divides into feveral branches, full of leaves, indented and pointed. At the extremity of each bough, appear radiated flowers, round, compofed of feveral well formed leaves, which are of a yellow colour. The difk confifts of feveral flourifhes, divided into many parts. Thefe flowers have likewife crowns, compofed of half-flourifhes, placed in a cup, of one leaf. The Indian pink requires much the fame management as the female. balfam apple. The cold injures them very materially. This plant is very proper

in

in all the compartments of our parterres : but
they fhould not be placed among plants of
the fmaller fize, nor in the middle of beds ;
for, by fuch a fituation, the great beauty of
thefe pinks would be loft to the fpectator.

L U P I N E.

LUPINES confift of three forts; the great
blue, the fmall blue, and yellow flowering
fpecies.   They all bloffom in May and June.
The firft fort grows to about two feet high ;
and the two latter, about half the height of
the former.   They are a flower that is feen in
moft gardens ; and are remarkable for their
neatnefs of bloffom, and fimplicity of colouring.   The yellow fpecies poffeffes an
agrecable fcent, which is denied to the other
forts, that however are recompenfed, in general, with a greater brilliancy of colouring.

C O N V O L V U L U S.

THIS plant confifts of three fpecies, called
the major, minor, and the fcarlet flowering
kind.   The major has a flower of a rich
purple colour ; the minor difplays a flower of
a delicate hue, between a fky and a marazine
blue : this fpecies is fometimes variegated
                                                with

with the colours of yellow and white. The
fcarlet-flowering kind is diftinguifhed for
bearing a flower, of the colour from which
it derives its name. But that which moft
particularly characterifes the convolvulus, in
all its three fpecies, is the flower, confifting
of a fingle leaf, which is a remarkable inftance of the variety nature difplays in every
part of the creation, when contrafted with
the ranunculus, and other flowers that are
compofed of fuch a multitude of leaves.
The convolvulus blows from June until Auguft ; and, as a picture of humility, creeps
upon the ground.

ASPHODEL.

THIS plant, from its appearance while
blooming, being fimilar to a royal fpear, is
called in Latin, *haftula regia*, i. e. king's
fpear. The ftem of the afphodel is three
feet high. In the middle of it grow, up to
the top, a great number of fingle flowers,
each divided into five parts. It thrives in
every fort of foil ; is multiplied more by
roots than feed ; and, if well watered, will
afford moft beautiful flowers. The afphodel
is confidered as a great ornament to a border, or any other part of a garden, where
dwarfs, or tall flowers, are raifed. It fhould
be fet three inches deep, and a fpan diftance

from

from each other, or from whatever flowers may be in the same compartments.

FOXGLOVE.

IS a large flower, resembling a thimble worn on the finger: from the root grows a stalk, two, and sometimes three feet high; and is hairy, and of a reddish colour: the leaves are oblong, and pointed at the end; covered with a little hair; indented on the edges: the outside is a brownish green, and the inside of a silvery white. On one side of the chief stem sprout several footstalks, which support single flowers that are wide at top, and are cut into two lines: their colour is generally purple, although they have sometimes a mixture of hues. In the middle of the cup is a chive, which adheres to the hind part of the flower. A light foil agrees best with this plant. The seed being very small, should be thinly sown in September. Foxgloves flower in June. Being tall plants, they are only adapted for the borders of beds, where the larger species of flowers are set or planted.

HEARTS

HEART'S EASE.

THIS flower, by the Latins, is called *viola tricolor*, from being adorned with three colours. It bears stems which have a tendency to creep along the ground; and are full of leaves, and rather oblong: the stems branch into boughs; at the top of which grow the flowers, which are placed under the species of violets, compofed of five leaves, from bearing a cup divided into five parts: each flower is white, blue and yellow-coloured. It is multiplied by feed fown in beds as thinly as poffible. When fufficiently raifed, it is removed into pots, where it makes a more agreeable appearance than it does in its native humble fituation, where it is loft and overlooked, like modeft merit, amid its greater and more fplendid neighbours.

AURICULA.

THIS flower has been the greateft pride of all gardeners. One root of it has fold for twenty guineas. Thefe flowers are indeed very delightful, both in fcent and beauty. They bloffom in April, and are in full bloom about the 20th of the fame month. The numerous variety of their flowers, are diftin- -guifhed

guifhed by the names and titles of eminent and exalted characters: thus, it has been not unaptly obferved, that, as auriculas increafed fo faft, and great men, if poffible, decreafed fafter, in a fhort time names of diftinction would be wanting to denote their differences. The goodnefs of an auricula confifts in a ftrong flower-ftem, fhort footftalks, large regular flowers, full, round, and white eyes; and that the flowers themfelves may be flat, not the leaft inclining to cup.

The culture being particular, we refer our readers to Bradley's new improvements in gardening and planting.

VIOLET.

THE violet produces, from its root, tufts of leaves almoft round, indented on the edges, and of a beautiful green. In the middle of thefe leaves grow the flowers; confifting of feveral irrregular lips, fhaped like a butterfly: the two uppermoft refemble a ftand; and thofe on the fide are like wings; and the two lowermoft are formed like a little bark. Thus curioufly formed, it has been equally the pride of the peafant, prince, and poet. It is one of the moft early beauties with which Flora prefents reviving nature. It grows in any fort of ground, and is particularly pleaf-

ing

ing upon the borders of fmall gardens. The flower is agreeable to the fmell as to the fight; which has caufed it to be fo univerfal a favourite. It fhould be replanted every three years, and kept from weeds, which is the chief trouble the culture of the violet requires. The double violet is only that which is raifed in our gardens.

PINK.

THIS plant fhoots long, ftrait, thick, hard leaves of a blueifh green. In the middle rifes the ftem, long, round, and jointed at a certain diftance : on the top of this the flowers grow, confifting of feveral variegated leaves, fupported by a hollow membranous cup. Such is reckoned the beauty of this flower, that it has been the firft ftudy of the moft emnient gardeners, to raife them in the greateft perfection. Volumes have been written on their cultivation; and, as the flower is fo well known, we fhall only add, that pinks are fet indifferently, either in open ground, upon beds, in earthen pots, or in tubs, in autumn, or the month of March. They are one of the chief ornaments of all gardens : and are remarkable for the variety, beauty, and excellence of the flower.

AU-

AUSTRIAN ROSE.

THIS plant has, like other rofes, a prickly ftalk, which is garnifhed with winged leaves of an oval form, and their lobes fawed. The flower confifts of petals that are indented at the top, and which have one fide red and the other yellow. It being a fhrub, it may be propagated from the fuckers that grow from the roots or from the offsets, either in fpring or autumn. It bloffoms during the months of July and Auguft. Although this flower is much cultivated, yet Miller obferves, that it is only an accidental variety of the rofe confidered as a genus. Among the many fpecies of rofes, this is cultivated as one of the moft valuable embellifhments of a fhrubbery.

HELLEBORE.

GROWS wild in Italy, Auftria, and Lombardy. It thrives beft on high fituations. It has a plain ftalk, ungarnifhed with leaves, until it produces the bloffom on its fummit: the flower is yellow, and compofed of five or more petals. The root is fibrous. This plant fhould be propagated by offsets, and the roots fhould be taken out of the ground,

and

and tranfplanted. When their leaves decay, which is generally from the beginning of June to October, the roots fhould be planted in fmall clufters, in order to improve the appearance of their blofloms. If planted alternately with fnowdrops, their effect will be the more agreeable, as they flower about the fame time.

I R I S.

THE bulbous iris fhoots forth a ftem, formed of long, broad leaves, that are foft, and of a pale green colour. In the middle grows a ftalk which bears, on its top, a fingle-leafed flower divided into fix parts; and, in the centre of the flower, is a chive of three leaves arched. Their flowers are either white, yellow, blue, red or afh colour, and are moft beautiful in appearance. They are multiplied both by their feed, and by bulbs. When the feed is to be fown, it fhould be gathered in July, and preferved until September, before it is committed to the foil; and whatever colour the feed is, you may expect to have a flower arife from it of the fame hue, which is a circumftance peculiar to the iris, and may account for its name, which is derived from a Greek word fignifying to foretell or prefage; for the feed thus foretells the colour of the flower.

NASTUR-

NASTURTIUM.

THE NASTURTIUM INDICUM, or Indian
creſſes, are of two ſorts ; one large, and the
other ſmall. The large ſort is known by the
name of monk's hood : it has flowers, variegated with yellow and ſcarlet : they run
upon the ground, and blow from May to
September. This plant is raiſed with little
care. The ſeed, being large, is ſown in ſeparate grains, at four inches diſtant from each
other. The flowers of monk's hood grow
upon ſmall reddiſh ſtalks, and are compoſed
of ſeveral irregular leaves. The ſtem is covered with leaves ; which are ſometimes
round, and ſometimes angular. The ſmall
ſort of naſturtium is frequently eaten as a
pickle; but the larger, which is monk's hood,
is conſidered as poiſonous.

HOLLYHOCKS.

CONSIST of ſeveral ſorts. They have a
large ſtem, that riſes about ſix feet high ;
which is decorated with flowers, in the ſame
manner as other flower plants are decorated
with leaves. The flower blends the delicacy of the poppy with the richneſs of the roſe.
The colours of theſe flowers are various ; as
the

the red, white, purple, and black. Although the ftems of the hollyhock are fo ftrong and large as to grow fix feet high, yet they wither every winter to the ground. Their feeds are fown in March, in the natural earth ; and, notwithftanding they lie not long in the ground, they produce no flowers until the next year. They may be tranfplanted about March or September. The time of flowering is in July and Auguft.

LILY OF THE VALEY.

MANY are furprifed that this plant fhould be called a lily, as the bloffom has not the leaft refemblance to that flower. Of this plant there are two forts ; the white and the large-leaved lily. The firft has a ftem a foot high, bearing three long, large, fmooth, green leaves : the ftem, from the middle upwards, is adorned with flowers almoft round, white, very fragrant, and faftened to a fmall fprig. The fecond only differs from the firft in having red flowers inclining to white, and not having fo agreeable a fcent. The lily of the valley is only multiplied by flips taken from the plant and roots. This plant, firft arifing in a valley, thrives no where fo well as in fhady places ; for which reafon, it is never fet in the walks, but in fome private part of

the

the garden, where it is reared for the fake of its flowers.

CROWN IMPERIAL.

THIS plant has a ſtem about two feet high, which is ſurrounded with long, pointed leaves, growing immediately from the root: the ſtem is likewiſe garniſhed with ſmall leaves, growing in pairs, without any foot-ſtalk. Upon the top of the ſtem is the flower, com-poſed of ſeveral green, upright leaves, that appear to grow from the germ of another. flower, formed of yellow inverted leaves, in a figure ſomewhat reſembling a turban: amid theſe leaves are ſeen ſtamina, with white anthera, which hang down in a grace-ful manner. The anthera reſemble dew-drops, falling from the filaments of the ſta-mina. The crown imperial is propagated from its bulbs, which ſhould be taken out of their mould in June, well cleaned, and care-fully ſtored till September; when they ſhould be replanted. It bloſſoms chiefly in March and April: during theſe months, its ſingu-lar beauty, and graceful dignity, form one of the chief ornaments of our moſt elegant gardens.

HYACINTH.

HYACINTH.

NEXT to thefe follows the hyacinth, with all its virgin beauties : there are fo many forts of them, and fo different in colour, that nature feems to have taken pleafure in forming them, and rendering them more admirable by variety. As we are noticing the more early flowers, we have to obferve, that the winter and fpring hyacinth is blue, and odoriferous. It is little, round, and of a fingle colour. Hyacinths, like many other flowers, are multiplied by feed. The bulbs that are produced from the feeds, bear no flowers until the fourth year. The greateft part of hyacinths delight in places that are expofed to the fun, and apart from other flowers. Like animals that herd together in flocks, hyacinths are, by nature, moft adapted to grow in clufters, by themfelves.

MARTAGON.

THE martagon, or mountain lily, confifts of feveral forts. The great martagon has a red flower, growing on a ftem between two and three feet high, without any footftalk. It is fmooth to the touch, and of a deep green : the flower is crooked, and bends down at

. the

the end of the stalk, which supports it from
falling. The plant may be set in any soil.
It must be planted a span deep in the earth,
and the same distance from any other flowers
which it accompanies. It is set among flowers of the larger size, or rather in middle of
borders, with flowers smaller than itself.
The martagon blooms in May. The bulbs
should not be removed before you intend to
transplant them. Being sooner affected with
heat than cold, the bulbs should be sheltered
from the sun with little layers of earth, or
preserved from summer heat by frequent wa-
terings.

SWEET PEA.

THIS plant is frequently introduced into
gardens from the sweetness of its scent, and
the delicate beauty of its flowers. It is generally set with another, called the painted
lady. The flower of the sweet pea is exactly
the same as the common pea blossom, except
being purple instead of white. The flower
of the painted lady is pink and white. They
are both raised from seed, which is sown a-
bout the time of the other pea. They blossom
mostly in July, and are no little decoration to those parts of a garden allotted for the
irregular beauties and simplicities of nature.
POPPY,

POPPY.

THE garden poppy has a ftalk about two feet high, which fupports a flower diftinguifhed for its delicate texture, beauty, and variety of colour, and its fomniferous odour : but although the flowers are fo agreeable in appearance, they are of fhort continuance. They fhould be fown in fpots, in order to afford an affemblage of colours, their variety of hue is fo well calculated to afford. This flower is faid to yield a fubftance which is generally fold by our apothecaries as opium. The Dutch wild poppy does not blow fo high as the former : The flowers are red and white ftriped, and bloom during the months of June, July, and Auguft.

MEZEREON.

THIS plant is of two forts : the red and white flowering. The red is very common in gardens; but the white mezereon is rather fcarce. They are both dwarfs, and feldom rife higher than three feet : their ftalks are ornamented with flowers fo early as January, when the air is perfumed with their agreeable odours. They remain a long time in bloffom, and are much adorned with the

beauty

beauty of their fruitage. The only mode
of propagating them, is by fowing their
feeds in March. This plant may be profit-
ably introduced into parterres, as a flow
flower, or in wildernefs works, for its delightful bloffoms. But they are adapted
chiefly for a winter garden.

HONEYSUCKLE.

IS a fhrub, which fhoots forth feveral
branches, that expand on every fide, and
fupport themfelves by twining round whatever is within their reach. At the knots of
the branches, the leaves grow in pairs, oppofite each other, at equal diftances : they
are foft, broad, pointed, green without, and
white within. At the end of the branches
the flowers grow, in the form of pipes,
bending in a manner fomewhat fimilar to a
crown. The peculiar form of the leaf, an
agreeable diverfity of colour, and the aromatic odour it difpenfes around the gardens
it decorates, render the honeyfuckle one of
the moft defirable appendages to every fpot
where the bounties of Flora are collected
for human delight.

St

Sᴛ. J O H N 's W O R T,

Gᴚ ROWS on a thin, leafy ſtalk, about a foot high. From the chief ſtem grow many branches, which are garniſhed with long, ſmall, pointed, and plain-edged leaves. On the top of each of the ſmaller branches, is a yellow flower, which greatly reſembles the daiſy, both in ſize and form. If reared in a green-houſe, this flower will bloſſom in March : but, if cultivated in a garden, the uſual time of flowering is in June ; when it may be gathered for medicinal purpoſes. St. John's wort is reared in moſt phyſic gardens, from its poſſeſſing qualities that greatly aſſiſt the cure of the jaundice : it is likewiſe a chief ingredient in that valuable balſam ſo well known by the name of Friar's balſam, or Turlington's drops.

T H E E N D.

Edward's

Dols. Cts.

Edwards on Religious Affections,	1	—
Oliphant's Sacramental Catechifm,	—	18
Scott's Leffons, the fixth American edition, with plates, containing above 400 pages,	—	80
Fifher's young Man's beft Companion, or American Inftructor,	—	75
Effays on various Subjects, principally defigned for young Ladies, By Mifs More,	—	37½
Sacred Drama, by ditto,	—	62½
Milton's Paradife loft, from the text of Dr. Newton,	1	—
———————— regained, from ditto,	—	25
Economy of Human Life,	—	20
The hiftory of Sandford and Merton; a new work highly recommended; complete, 2 vols. in 1.	——	88
The Iliad of Homer, tranflated from the Greek, by A. Pope,	1	—
Letters of certain Jews to Voltaire,	2	—
Gough's Arithmetic,	—	80
Offian's Poems,	2	—
Stuart's Philofophy of the Human Mind,	2	—
Workman's Gaughing,	1	37½
An Eftimate of the Religion of the Fafhionable World,	—	37½
Reid's Effays on the active and intellectual powers of Man, 2 vols.	4	—
Perrin's French Grammar,	—	80
An Effay on Punctuation,	—	40
Sheridan's English Grammar,	—	44

THE

American Accountant;

O R,

Schoolmasters' New Assistant.

Comprised in FOUR BOOKS.

Book I Containing Arithmetic of Whole Numbers,—Divers Denominations, and the Common Rules, to the end of the Double Rule of Three.

Book II. Fractions, Vulgar and Decimal.

Book III. Mercantile Arithmetic; or all the Rules necessary for forming a complete Accountant; methodically arranged and largely exemplified.

Book IV. Extractions, Progressions, &c. being the higher Rules of Arithmetic.

AND

Including all the Questions in the Philadelphia Edition of G O U G H, with many others. The Rules are either new, or those of that Treatise so compendized, as to be both brief and perfectly applicable.

The whole adapted to the Commerce of the United States ; and comprehending every Thing necessary to a complete practical Knowledge of the Science of Arithmetic.

───────

By BENJAMIN WORKMAN, A. M.

───────

The THIRD EDITION.

Revised and corrected by R. PATTERSON, A. M. Professor of Mathematics in the University of Pennsylvania.

Price Half a Dollar.

WILLIAM YOUNG,

Has on hand an extensive assortment of PA-
PERS, from the best manufacturers in
EUROPE, and from his manufactory on
BRANDYWINE.

WRITING AND PRINTING PAPERS, *viz.*

Imperial,
Super-royal,
Royal,
Medium,
Demy,
Thick Post, in folio,
Ditto, in 4to,
Extra large Folio Post,
Ditto, 4to,
Folio Post, wove,
Quarto, ditto,
Folio wove Post, lined,
Quarto, ditto, do.
Ditto, gilt, do.
Common size Folio Post,
Ditto, quarto, plain,

Folio & quarto Post, gilt,
Small Folio Post, plain,
Ditto, gilt,
Blossom Paper assorted,
Transparent Folio Post,
Superfine & comm. Foolsc.
Marbled Papers.

COARSE PAPERS.

London brown assorted,
Log-book paper,
Hatter's paper,
Stainer's paper,
Common brown,
Patent sheathing paper,
Bonnet boards,
Bookbinder's boards.

ALSO, a variety of other STATIONARY ARTICLES, *viz.*
Wedgwood and glass Philosophical ink-stands well assorted, pewter ink-chests of various sizes, round pewter
ink-stands, paper brass and polished leather ink-stands
for the pocket. shining sand and sand-boxes, pounce
and pounce-boxes, ink and ink-powder, black leather
and red morocco pocket-books with and without instruments of various sizes, counting-house and pocket
pen-knives of the best quality, ass-skin tablet and me-
morandum books, red and coloured wafers, common
size office ditto, quills from half a dollar to three
dollars per hundred, black lead pencils, mathematical
instruments, &c. &c.

All sorts and sizes of BLANK BOOKS ready made or made
to order, bank checks, blank bills of exchange, and
notes of hand executed in copper-plates, bills of lading,
manifests, seamen's articles and journals, &c. &c.

A well selected collection of miscellaneous books. ALSO,
of Greek, Latin, and English Classics, as are now in
use in the Colleges and Schools of the United States.

www.ingramcontent.com/pod-product-compliance
Lightning Source LLC
Chambersburg PA
CBHW021405210326
41599CB00011B/1014